Lecture Notes in Electrical Engineering

Volume 23

Lecture Notes in Electrical Engineering

Roland Beutler

Digital Terrestrial Broadcasting Networks

 Springer

Roland Beutler
Sdwestrundfunk
Stuttgart
Germany
Roland.Beutler@ swr.de

ISBN: 978-1-4419-3500-7   e-ISBN: 978-0-387-09635-3
DOI 10.1007/978-0-387-09635-3

Printed on acid-free paper

springer.com

To Kai and Chris

Contents

Preface

Frequency management and network planning of terrestrial digital broadcasting systems are both fields of activity that are certainly not part of natural sciences. They cannot even be classified as engineering science. Without doubt, communication science and electronic engineering constitute the basis for frequency management or network planning. However, none of these academic disciplines cover all aspects.

Typical frequency planning problems exhibit such a high level of complexity that very sophisticated mathematical methods have to be consulted in order to find solutions. Furthermore, a profound knowledge in physics is also required, since in any case, any transmission of information is governed by the Maxwell equations. This is independent from the way the transmission is actually accomplished, that is if either terrestrial transmitter networks are used, cable connections, or satellite links.

Combining methodologies from natural and engineering sciences is still not enough to fully covers the whole variety of different tasks planners are confronted with in the field of frequency and network planning. There are also technical aspects related to the operation of transmitter sites that are directly connected to economic questions. Moreover, every planning strategy must be embedded into a corresponding media political milieu.

Different countries have different approaches and strategies when it comes to providing broadcasting services. Therefore, different rules for regulation might be valid in different countries. However, since electromagnetic waves do not stop at national boundaries there is always an international administrative and political aspect of frequency management and network planning. Politics and economics impose important constraints on the planning process, which in the end very often leads to a situation that mathematical algorithms are not able to provide acceptable solutions. Under such conditions frequency plans or network designs are drawn up by negotiations between administrations of neighboring countries.

Frequency or network planner is not a job people are usually taking up intentionally, as they choose the focus of their academic studies, for example. In most cases pure chance plays a major role. Hence, it comes as no surprise that in the field of frequency and network planning people from nearly all technically oriented faculties can be found. There are engineers from communication science to electronics accompanied by mathematicians and last but not least a lot of physicists.

Very likely all of them share one common experience, namely the first step on the new ground called frequency or network planning proves to be rather weary. Basically, this is connected to the fact that, in particular, network planning is an activity where a lead in knowledge gives a direct competitive edge. Consequently, successful planning approaches are published only to the extend that an ascendancy in the market of network providers is not lost by giving away secret expertise. Such an attitude is very common, as can be seen, for example, from the situation on the radio or TV receiver market. All manufactures need to build their products in accordance with well-known published standards. Nevertheless, there are good and bad receivers on the market to be bought at nearly the same price.

Besides the lack of officially published documentation, the highly practically oriented working methods of the planners have to be considered a real obstacle for any novice. Many of the ideas will never be published in articles. Most of the work is carried out by project groups developing new concepts that are not documented very elaborately. Once the new ideas have been put into practice, most of the presentations and manuscripts vanish again.

Clearly, the results of international frequency planning conferences are memorized very carefully and detailed. The documentation prepared under such circumstances must contain precise directions as to how to make use of the planning results. But at the same time they represent some kind of a very compressed, condensed information. For a newcomer this might lead to nearly the same level of forlornness as having no documentation at all.

The background and the experience of the author during his confrontation with frequency and network planning gave umbrage to his first book on frequency and network planning for digital terrestrial broadcasting that was published in 2004. It was meant to give an insight into the problems and thereby emerging strategies to find solutions for them. This book again addresses the field but with a slightly different focus. The first book introduced basic concepts and tried to illustrate them by putting forward several detailed, but artificial examples. This approach was chosen because it is quite obvious in view of the manifold of different tasks under the umbrella of frequency and network planning that no exhaustive presentation of the whole field can be provided. Rather, it was intended to sketch the principal ideas.

In contrast, in addition to giving an idea about the characteristics of typical planning problems this book discusses the results of important international planning conferences that define the constraints frequency and network planning are subject to.

As a consequence, this book starts with a presentation of several relevant digital terrestrial broadcasting systems. Their different characteristics are described to some extent. Then, an overview about the structure of international spectrum management is given. The major organizations and bodies are presented together with their tasks. Chapter 4 lays the foundations of coverage prediction, which is the crucial element for both frequency management and in particular network planning. The characteristics of the terrestrial radio channel are addressed, which are the starting point for the development of wave propagation models. The chapter ends with a presentation of methods for coverage assessment. Frequency planning basics follow in the next chapter. All relevant quantities and parameters are introduced and explained in detail. Much emphasis is put on mathematical algorithms and strategies to create flexible and robust planning tools. Chapter 6 deals with network planning for digital terrestrial broadcasting. Again, the focus is on the presentation of mathematical methods that are suitable in order to optimize network structures. In particular, the application of stochastic optimization algorithms is discussed. These methods open the door to cope with very different planning scenarios taking into account different aspects including network costs.

Starting with Chapter 7, the character of the book becomes less theoretical as those international frequency planning conferences are described, which are relevant in relation to digital terrestrial broadcasting. The book ends with an outlook to future developments, in particular the perspectives of terrestrial broadcasting in general.

In each section of the numerous references other documents are given. However, the selection of documents referenced in this book has to be considered as a subjective subset of what might be relevant. Nevertheless, it gives first indications and hints for further reading.

Chapter 1

Introduction

The availability and the unbarred access to any kind of resources are the driving forces for all civilizations. Food, clean water, and clean air are all essential, and so is energy in all its aspects. Usually, the term resource is automatically linked up with material resources. In recent times, however, immaterial resources are getting more and more importance. In this age of globalization, states possessing only few or no natural resources are becoming increasingly dependent on the fact that their working population has sufficient technical or economical knowledge and know-how. This might allow them to compensate for the lack of natural resources.

Services of very different kinds take the place of simple processing of raw materials or the agricultural production of food. Of course, the generation of knowledge and its provision presumes the existence of adequate communication possibilities. The so-called information age, as a consequence, promotes and puts into practice a general digitization of information and all types of information carriers on every level in society. There is basically no part of our daily life that is not subjected to dramatic change. Without the help of computers, the synonym of digitization, countries around the globe could no longer sustain public life. In particular in those fields of activity that are to be considered essential like public services and economy computers have become an indispensable tool. Even though they are omnipresent, it seems that this development has just begun. Where it will lead to is far from clear. More and more tasks are currently deputed onto computers. The performance of state-of-art computer technology today is already breathtaking

R. Beutler, *Digital Terrestrial Broadcasting Networks*,
DOI 10.1007/978-0-387-09635-3_1, © Springer Science+Business Media, LLC 2008

and every day new applications are presented showing new, so far, not reached functionality.

The amazing changes that are taking place today can be seen no where better than when looking at the incredible development of the Internet. It took less than a decade to initiate a far-reaching social transformation, which compares to a phase transition in physics. The networking of computers across the whole planet implies completely new economic and political structures whose profound consequences today are still to be discovered. By using the Internet, people are able to communicate nearly without limitations in space and time. They can share any type of information by employing cable connections, satellite links, or radio transmissions. When establishing a point-to-point connection across a long distances, very likely, all three communication types are involved.

A common feature of all these possibilities is that they make use of electromagnetic waves as the medium for the transmission of information. Electromagnetic spectrum is an immaterial resource. It is limited but in contrast to petrol the reserve will never run short. In addition, it is not subject to a reduction of quality like, for example, air, which gets polluted by harmful emissions. Electromagnetic spectrum is available everywhere, either on ground, under water, or in outer space. Of course, not all the parts of the electromagnetic spectrum are equally well suited for the transmission of information. Figure 1.1 sketches the different regimes, which according to their characteristics can be used for different applications.

One of the central problems that the information society has to overcome in the future is to provide transmission systems, which can cope with the permanently increasing demand of people for different types of communication. Clearly, the development and introduction of such systems is limited by the scarcity of appropriate electromagnetic spectrum. Consequently, if the amount of spectrum cannot be increased, forthcoming generations of telecommunication systems must be designed to allow for a highly efficient usage of spectrum. Depending on the type of communication, different approaches might be adapted.

Basically, there are three different communication levels to be distinguished. If one person tries to establish a communication link with a single other person, this is simply called a point-to-point connection. In addition, there exist point-to-multipoint and broadcasting scenarios. Point-to-multipoint is typically encountered in computer networks,

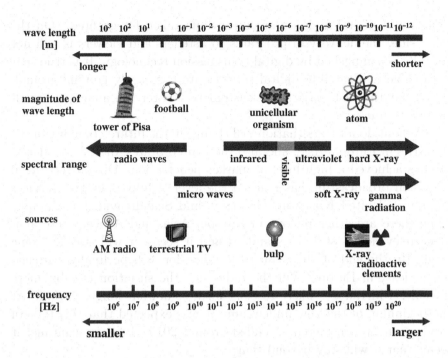

Figure 1.1: Overview of the electromagnetic spectrum.

for example, when several people connect to the same Web server to download exactly the same content. The number of users thereby tends to be from one to some hundred persons at the same time. Broadcasting, in contrast, means to distribute the same information in principle, to an unlimited number of customers at the same time. It is evident that all of these communication types give rise to completely different requirements for the underlying system design. It should be noted, however, that this does not prohibit mutual compatibility and transparency. Indeed, it is crucial that interfaces have to be developed that allow for a seamless link between different communication platforms. Furthermore, it has become clear in recent years that in the future it will be mandatory for transmission systems to possess the ability of dynamic operation mode reconfiguration in order to allow for a very efficient usage of the available data capacities.

The profound structural change of societies around the world started to seize both radio and television at the beginning of the 1990s. On one side, the entire production chain of radio and television programs was switched to digital. Digital cameras, digital editing of audio and video signals and digital archives are meanwhile taken for granted. On the other side, the delivery of programs to listeners and viewers is still not yet fully accomplished by digital transmission technology. It is true that both cable and satellite digital services are constantly gaining ground. But so far the vast majority of customers still rely on analogue broadcasting technology.

Without doubt the situation will change in the future. In many of the countries in Europe, Asia, and also the United States, digital broadcasting has achieved a high level of market penetration. Digital terrestrial television broadcasting has reached the level of fully operative services already in United Kingdom, Germany, and Scandinavia, for example, while there are other regions in Europe, Asia, and Africa where analogue television is still the standard and it seems it will stay for some time. By the end of 2012, analogue television will be finally switched off, at least in Europe. On the radio side the situation is a bit more difficult. FM radio is likely to exist for a somewhat longer period. At the beginning of the new millennium, it was expected that this type of service can be considered as ended around 2015. In the meantime, it seems that it will stay beyond that.

Before the all digital terrestrial broadcasting scenario can become a reality a lot of obstacles must be removed. A rigorous switch-over from analogue to digital transmission will probably not be possible. It seems that a more or less extended so-called simulcast period will be necessary for the introduction of digital terrestrial radio and television. Simulcast means the simultaneous broadcast of one particular program by both analogue and digital means. The question how the costs naturally arising by such an simultaneous operation can be compensated for in an acceptable way cannot be answered yet. At least currently, the public opinion does not support the idea of longer lasting simulcast phases since nobody is willing to pay for that. On the other side, switching off all analogue transmitters at one chosen date would certainly create a lot of political and social concern and tension, depending on the number of customers and receivers in the market who are still actively using it. There seems to be no chance for any acceptance of such an idea as a general rule and consequently, it could not be enforced.

Hence, strategies for the introduction of digital terrestrial broadcasting systems have to be developed that allow for a smooth transition. However, it is evident that this also will lead to a great variety of problems to be solved. In the first place, spectrum scarcity has to be mentioned here. A brief look on analogue television in Germany might clarify this. In 2000, analogue terrestrial television occupied the entire frequency range of Band IV and V in the UHF (Ultra High Frequency) range. Only on a very few numbers of occasions could new TV channels be found at a particular location, if a gap in the existing coverage had to be closed. In most cases, it was not possible to improve the coverage with analogue TV programmes. Since digital terrestrial TV was to be implemented in the same frequency range, it was unavoidable that its introduction had to be accompanied by switch-off or rearrangement of analogue transmitter sites.

So far digital terrestrial radio and television have been mentioned without specifying any system characteristics. Nevertheless, it has been tacitly implied that in Europe the two standardized systems "Digital Audio Broadcast" (T-DAB) [ETS97a] and "Digital Video Broadcast" (DVB-T) [ETS97b] are to be introduced and hence treated here in the first place. Both systems have been designed for several different transmission environments. The letter "T" in the abbreviations specifies the terrestrial variant of both systems. There exist standards for both cable and satellite transmission systems. The position of the letter "T" is purely historically based and has no deeper meaning.

In particular, in the United States of America there exist competing terrestrial radio and television systems, which are also designed for terrestrial broadcasting purposes (see e.g., [ITU01b]). It seems as if at least in Europe and wider parts of Asia and Africa T-DAB and DVB-T will be the favorite digital terrestrial broadcasting standard. However, the race is not yet finished. In Japan, a system has been put forward that combines radio, television, and telecommunication under one single technical roof. Especially, from the point of efficient usage of the spectrum such an approach might seem to be very attractive (see e.g., [ITU02]).

Efficient usage of the available spectrum is also one of the characteristic features of both T-DAB and DVB-T. A relatively large bandwidth occupied by the broadcast signal combined with sophisticated technical measures gives rise to very robust transmission systems. In particular, T-DAB has been designed and optimized for mobile reception. Portable or fixed reception are clearly possible without problems as well. In the DVB-T case, the latter reception modes were the major design targets of the system. But even for mobile reception, significant improvements could be achieved, which are due to more efficient antennas and better design and performance of the electronic components of the receivers.

Analogue radio and television systems require a large amount of spectrum to provide a wide area coverage with one dedicated content.

In order to serve an area in the order of 200 × 200 km with only one FM radio or analogue television program, several frequencies or channels are needed. A typical FM high-power transmitter has an output of about 100 kW. In case there is no other transmitter around, using the same FM frequency in an area of up to 100 km radius could be served, depending on the topography.

In practice, such an isolated operation of a transmitter is never met anywhere. As a consequence, there are extended areas where the signals from different co-frequency transmitters superimpose at the location of the receiver. Analogue radio or television receivers cannot cope very well with such a superposition of different contributions from different sources. Usually, the reception will then fail. Basically, this leads to a reduction of the theoretically achievable service area. Moreover, if two transmitters using the same frequency would broadcast the same content, an FM radio or an analogue television receiver would have problems decoding the information properly. This is the reason why within an extended area of the dimensions just mentioned, 10 or more high-power transmitters using different frequencies are utilized to provide analogue radio or television programs. Under such conditions they have a typical mean inter-transmitter distance of about 50 km. The important issue, however, is that each transmitter needs to use its own frequency well separated from all other frequencies.

Both T-DAB and DVB-T are based on techniques that differ completely from those employed in their analogue counterparts. First of all, several programs are bundled into so-called multiplexes and thus broadcast simultaneously. This possibility is a direct consequence of the system design. Both T-DAB and DVB-T employ a large number of individual high-frequency carriers distributed across the spectrum range used by the systems to carry the information. This is the basis to compensate effectively for service degradations due to critical wave propagation conditions.

As mentioned, the spectrum being at the disposal for a T-DAB or a DVB-T transmission is exploited to transmit several programs at the same time. Audio or video data together with additional information are combined to build a multiplex, which with the help of modern modulation schemes is transmitted by the set of high-frequency carriers. In the case of T-DAB, six to eight audio programs in nearly CD quality per multiplex is the rule, while for DVB-T two to four PAL equivalent programs can be broadcast within one multiplex. Clearly, if more programs than that are to be provided within a particular area, additional frequency resources are needed in order to establish and broadcast additional multiplexes.

The most important consequence of the system design for frequency and network planning of T-DAB and DVB-T is the possibility to implement the systems in terms of single frequency networks (SFN). SFN configuration means that all transmitters constituting the network within a specified coverage area are employing the same spectrum range in order to broadcast the same content. Clearly, also with digital signals superposition of different signal contributions is encountered. This is just physics and is naturally not connected to the characteristics of the broadcast signals. However, even though there is definitely a negative impact for certain parts of the used spectrum range of the signal due to the superposition of several signal contributions, it turns out that on an average superposition will lead to better reception. It must be emphasized, nevertheless, that this statement is true only as long as the spread in the times of arrival of the signals at the receiver location is not too large. The details of this particular system feature will be in the focus of the discussion later in this book extensively.

Since the propagation of electromagnetic waves is governed by basic physical laws it does not naturally terminate at national or international borders. Therefore, the assignment of frequencies onto coverage areas is in principle an issue of international nature. It has to be dealt with in accordance with all affected foreign neighbors. There are several international organizations who organize and manage the usage of the entire electromagnetic spectrum. The most import body certainly is the International Telecommunications Union (ITU). For Europe the Conference Européenne des Administration des Postes et des Télécommunications (CEPT) plays a major role as well. Chapter 3 will give an overview about all relevant international organizations and bodies.

In 1995, a CEPT conference was held at Wiesbaden, Germany, in order to establish the access to spectrum for T-DAB and to set up a corresponding frequency plan. Both in the VHF range and in the L-Band certain spectrum ranges have been allotted that gave the possibility to allow every country in Europe the implementation of two nation-wide coverages with T-DAB. It turned out that a vast number of different national and political constraints had to be taken into account explicitly. This led to a frequency plan that formally is full of incompatibilities. Fortunately, most of these problems could be solved by bi- or multilateral coordination. The reason for the incompatibilities lies in the very different spectrum usage across Europe. There were a lot of countries where the originally envisaged spectrum ranges could not be used because other telecommunication services were already occupying them. In view of these obstacles, it is evident that the assembly of a detailed frequency plan was only possible by the massive use of computer power and the exploitation of highly sophisticated mathematical methods.

The solution to the problem of finding a set of mutually compatible frequencies for well-defined coverage areas seems to be very easy and straightforward at first glance. Indeed, a more closer look reveals the extremely high complexity of the underlying optimization problem. Already the simplest task, namely just to find a set of frequency assignments for some areas so that there are no compatibility problems and at the same time the use of spectrum is minimized falls into the class of mathematical problems that are labeled "NP-hard." A typical feature of such types of problems is that if one would try to find the one and only optimal solution by checking every possible solution, this would lead to a computational effort, which increases incredibly fast as the characteristic system size increases. For the case of a frequency assignment this means the computation time blows up faster than exponentially as a function of the number of considered service areas. In the case of typical, realistic numbers of service areas to be included in the investigation, it proves that it is no longer technically feasible to find a solution by pure trial and error. This is true irrespective of any future performance boost of the then available computer technology.

What is left then as the only practical way forward is to change the search target from the one and only optimal solution to a nearly optimal or at least a very good solution. During the last decades there has been tremendous progress providing adequate methods thereto. Nowadays, there exist a vast number of different means to find satisfying solutions for frequency assignment problems in the sense just mentioned. A very elaborate list of references to relevant work can be found in [ZIB01]. It has to be noted that most of the articles listed there refer to mobile telecommunication and not to broadcasting. However, the principal ideas can be applied to the field of broadcasting as well, since the employed concepts still remain valid.

After the end of any frequency planning conference, all participants hopefully are in possession of the rights to use particular frequencies within their defined coverage areas. This is the time when network providers can seriously start to think about the implementation of their networks. The question what type of coverage targets can be realized must be addressed. Typical examples would be wide area coverage or coverage according to the population density. Besides, the type of reception that is focused on determines the details of the network implementation as well. Very different efforts have to be made if fixed, portable, or mobile reception is to be aimed at.

Whatever the coverage intentions are, the characteristics of the T-DAB or DVB-T networks have to be specified in all details. This means answers to several important questions have to be found. How many transmitters are actually needed? At which locations should they be preferably situated? What are the effective output powers that are necessary to reach a certain coverage goal? In order to find an answer to these questions a model for the whole broadcasting network has to be developed. Clearly, wave propagation issues have to be included as well. A high-power transmitter on a high mountain will certainly have a completely different impact on the coverage than a low-power transmitter somewhere in the middle of a big city with large buildings or even skyscrapers nearby. The freedom to adjust the number of transmitters in the network and their individual characteristic properties seems to be unrestricted, at least in terms of a theoretical planning model. It is obvious that once again the planner is confronted with an optimization problem possessing an enormous number of potential solutions. And once again the task is to find a solution that compared to predefined requirements is (nearly) optimal.

The present book tries to give insight into the manifold of mathematical and practical problems of frequency assignment and network planning for digital terrestrial broadcasting systems. Frequency assignment and network planning are dealt with in two different chapters where basic aspects are presented. Even though a subdivision into two chapters might suggest that both fields of planning are independent, this is definitely not true. Frequency planning for T-DAB and DVB-T is usually based on the so-called allotment planning approach. It will be presented later in full detail. For the time being, it is sufficient to mention that this means to assign a frequency to a given geographical area to be used there without specifying the details of a corresponding transmitter network implementation. Seen from that perspective, network planning is always the second step.

However, it is important to understand that the generation of a frequency plan has to take into account the characteristics of transmitter networks that will be put into operation and the envisaged services areas. The topographical conditions encountered in the planning area need to be accounted for as well. All these conditions have to modeled appropriately for the frequency planning process in order to generate a frequency plan. This allows to subsequently implement transmitter networks that are reasonable from both a technical and an economical point of view. Clearly, the implemented networks then need to obey the restrictions associated with the usage of entries in the corresponding frequency plan. So, both frequency assignment and network planning are inseparably linked.

Chapter 2

Digital Terrestrial Broadcasting Systems

The success story of radio and television was originally based on terrestrial transmission. At a suitable location a transmitter was erected that broadcast radio and TV signals, using corresponding equipment and antennas. The listener or viewer was expected to maintain a certain reception effort. In the case of television, he or she was to implement an antenna on top of the roof of the house having a sufficiently high directivity. Most of the stationary radio receivers were fed by correspondingly adopted antenna systems.

At the end of the twentieth century the supremacy of terrestrial broadcasting over other distribution paths was definitely lost. At least this is true for television. According to market studies, in 2000 only 12–15 % of German households, for example, were still relying on terrestrial broadcasting as their primary reception method for television. One of the reasons certainly was the lack of a sufficiently high number of attractive programs that could be provided by terrestrial means. This, however, was directly linked to the scarcity of corresponding frequencies.

For radio the situation is different in several aspects. Terrestrial broadcasting still is the most important way to deliver audio programs to the listeners. Nevertheless, also in the case of radio an irreversible change is taking place. Since the introduction of CDs as carriers for music, people started to get used to the high-quality standards provided by that digital audio system. FM radio is not able to offer an equivalent service quality.

R. Beutler, *Digital Terrestrial Broadcasting Networks*,
DOI 10.1007/978-0-387-09635-3_2, © Springer Science+Business Media, LLC 2008

Another difference between radio and television is that the main reception environment for radio is not fixed reception, but portable or mobile reception. It should be noted that FM radio has not been designed for mobile reception in a vehicle. Originally, only fixed reception using a roof top antenna as in the case of analogue television was foreseen. Hence, more or less serious perturbations have to be accepted under portable or mobile reception conditions.

Basically, FM radio and analogue television suffer common problems. As in the case of TV, there is not enough spectrum to satisfy all the demands that public and private broadcasters would wish to satisfy. In recent years, the former very strict rules applied for coordination of FM transmissions [ITU98], were very gradually softened in order to find new frequencies at all. The current frequency usage situation is very tight. In most cases, the benefit gained by bringing a new FM frequency on air in some area has to be paid for by proportionally high level of interference at other locations. In other words, the FM spectrum has been squeezed out to the maximum.

All these facts triggered the development of the new digital terrestrial broadcasting systems T-DAB and DVB-T. Primary objectives were a resource saving usage of radio frequencies, high-transmission quality, and a large enough data capacity to allow for a sufficient number of offered programs. In the case of radio, the possibility for mobile reception even at high velocities was also one of the objectives. The target of the latter requirement were in particular vehicles moving on highways and also high-velocity trains.

T-DAB and DVB-T quite naturally exhibit differences since they are optimized with respect to the transmission of audio or video data, respectively. Nevertheless, they have several technical features in common. Both systems employ a multicarrier technology and in both cases psychoaccoustic or psychovisual insights are exploited to significantly reduce the amount of data that needs to be transmitted in order to maintain a certain level of quality of the audio or video signals. Furthermore, the transmission is protected against perturbation by the application of sophisticated error protection mechanisms.

In the sequel, both systems will be described very briefly. The focus will lie on those aspects of the systems that are essential in connection with questions arising in the context of frequency assignment problems and network planning. For a very detailed technical description it is referred to the standardization documents in [ETS97a] and [ETS97b] as well as to the textbooks in [Lau96] and [Rei01].

Both T-DAB and DVB-T mark just the first step into the digital terrestrial broadcasting era. They have been developed almost two decades ago. Clearly, the most advanced and appropriate technologies existing at that time have been incorporated. Even if at the moment both are going to be introduced or already have been introduced on a large scale across Europe, seen from a present perspective their technologies might seem to be outdated already. Technology has dramatically evolved since the T-DAB and DVB-T standards have been issued. This is reflected in the fact that enhancements of T-DAB and DVB-T have been introduced. Furthermore, entirely new systems such as MediaFlo have entered the stage of digital terrestrial broadcasting.

T-DAB has mutated into T-DMB, which is the abbreviation of "Digital Multimedia Broadcasting" and represents an extension of T-DAB capable to provide video data to a certain level as well. In the same manner, DVB-T has been further developed into DVB-H where the "H" stands for "handheld." Clearly, the idea is to broadcast television or video content to handheld devices similar to mobile phones.

Besides T-DAB there exists another digital transmission system, which is suitable for terrestrial radio broadcasting. It is called Digital Radio Mondiale (DRM) [ITU01b]. It has been designed to substitute the analogue systems in the short- and medium-wave regions. DRM is a multicarrier system, too. But in contrast to T-DAB or DVB-T, only a very small bandwidth is used, which is nevertheless very well adapted to short- and medium-wave transmission purposes. The DRM standard [ETS05] is currently undergoing the process of extension in order to allow the operation of the systems in the VHF bands I and II as well. Band II basically corresponds to the spectrum range where analogue FM transmissions take place. Therefore, the extended version of DRM for which the name DRM+ has been coined might also be a candidate system to substitute FM transmissions in the future. All these systems will be discussed here to some extent. For more information references to corresponding literature are always given.

Mainly in Japan another digital terrestrial broadcasting system is being introduced. It is called Integrated Services Digital Broadcasting-Terrestrial (ISDB-T) and combines system characteristics of DVB-T and T-DAB (see e.g., [ITU02]). Due to the close technical resemblance to T-DAB/DVB-T the Japan system will not dealt with here, too.

2.1   COFDM Modulation

In contrast to cable- or satellite-based distribution of radio and television programs, terrestrial transmission has to cope with the negative impacts of multipath propagation conditions. Typically, radio and television programs are broadcast from a transmitter, located at an elevated geographical site. Such a site is usually chosen in order to cover a large area. The signals are broadcast with a constant radiation power. Usually, the transmitting antenna gives rise to an angular radiation pattern, which means that in some directions more power will be radiated than in others. On its way from the transmitter to the point of reception the broadcast signal can undergo reflection or diffraction caused by obstacles like mountains, hills, or buildings. Thus, in general, at the receiver site not only the signal arriving from the transmitter along the direct path is received, but also contributions caused by the physical effects mentioned. These additional signals arrive under different angles, with different amplitudes, and at different times than the direct signal. Figure 2.1 sketches a typical transmission scenario in the case of mobile reception.

Physically, the linear superposition of the different individual signal contributions generates a characteristic interference pattern in the vicinity of the point of reception. Constructive interference leads to points where the resulting amplitude takes a maximum value, whereas in other locations due to destructive interference the amplitude is minimal. The field pattern depends on all details of the received signals, that is the number of contributing waves, their individual amplitudes, their angle of incidence, and last but not least the relative times of arrival. And ultimately, all that depends in detail on the frequency of the electromagnetic waves.

In the first place, the reception quality for a particular transmission system is determined by the field strength that is delivered at the point of reception. Therefore, it is possible to improve a deficient stationary reception quality simply by moving the receiver to a place where the field strength is larger. Due to the interference pattern in the vicinity of the receiver, the field strength varies across a distance in the order of

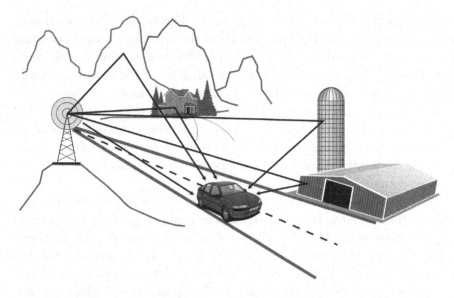

Figure 2.1: Typical multipath reception scenario for terrestrial transmission.

the used wave length. A very well-known example for such a situation is mobile FM reception. A driver moving in a vehicle and listening to a dedicated FM program might experience a reduction of service quality when coming to stop, for example, in front of traffic lights. The position of the car might coincide with the location of a minimum of the field strength caused by the destructive interference of several signals. Moving the vehicle just about 1 m might clarify the reception again.

However, sufficient field strength is not the only criterion for acceptable reception quality. Different signals arriving at different times might lead to a large field strength level due to constructive interference. However, the phase relations between the different signals are such that still the audio or video signals are distorted. In the case of analogue television multipath propagation typically leads to so-called ghost images. This corresponds to the appearance of laterally shifted weak pictures in addition to the proper TV picture.

In addition to the problems that arise from multipath environments, mobile reception imposes further critical requirements to be met. While moving the receiver passes through the interference pattern of the electromagnetic waves. This is equivalent to a dynamic change of the usable field strength level on a short time scale. Furthermore, the motion of the receiver gives rise to Doppler shifts of the arriving waves. Depending on the angle of incidence, relative to the direction of motion, the shifts are varying. As a consequence, the frequencies of the broadcast signal and the frequency the receiver is tuned to do not match anymore. The effect is frequency dependent in the sense that higher frequencies will suffer a larger Doppler shift. Also, increasing velocity gives rise to larger mismatch. Usually, broadcasting systems are able to cope with Doppler shifts up to a critical value, which depends on the characteristics of the system design.

Multipath propagation affects the performance of broadcasting systems without doubt. Its negative influence cannot not only be observed in the time-domain, but also when looking at the spectrum of the transmitted signals. Superposition of several waves leads to periodic notches in the spectrum. The extent of these degradations basically depends on the so-called delay spread Δt. This parameter describes the difference of the times of arrival between the first and the last incoming signal. It is linked to the coherence bandwidth B_C of the radio channel according to

$$B_C = \frac{1}{\Delta t} \; . \tag{2.1}$$

The coherence bandwidth corresponds to the approximate maximum bandwidth or frequency interval over which two frequencies of a signal are likely to experience comparable or correlated degradation by multipath effects. In order to assess the degree of degradation the signal undergoes, B_C has to be compared to the bandwidth B the transmission system employs. The case

$$B > B_C \tag{2.2}$$

gives rise to so-called frequency selective fading, which means that different portions of the signal bandwidth separated roughly by B_C experience decorrelated fading degradation. The other case, that is $B < B_C$ is called flat fading and describes that all frequency components of the signal will experience the same magnitude of fading.

Figure 2.2 is provided to give a graphic visualization of the impact of multipath propagation on the spectrum of the received signal. Two cases for different values of the time spread Δt are shown. The pictures have been generated by using an arbitrary broadband signal $x(t)$, which

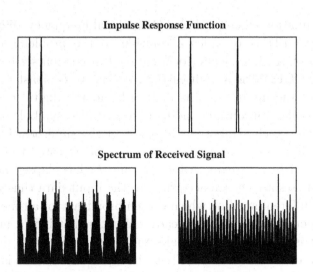

Figure 2.2: Impact of multipath propagation on the spectrum.

has been manipulated according to the impulse response functions in the upper part of Figure 2.2 to derive a fictitious received signal $r(t)$, that is

$$r(t) = x(t) + x(t - \Delta t) . \tag{2.3}$$

Then, a simple FFT has been applied to calculate the spectrum of $r(t)$.

If Δt is small, as on the left part of Figure 2.2, the spectrum is distorted throughout a larger frequency range. In the other case, the effect is not that pronounced. The distortion becomes more located. However, both cases correspond to frequency selective fading.

The impact of multipath environments is different for systems using a small or large bandwidth. The smaller the used bandwidth of the transmission system is the more dramatic the consequences of multipath transmission can become. Total failure cannot be excluded under some circumstances. In digital terrestrial broadcasting delay spreads in the order of 10^{-4} s are quite common. This is connected to the typical separation of transmitters in a corresponding network implementation (see Chapter 6). So, in the case of two nearly equally strong signals every 2 kHz the spectrum shows the characteristic fading depicted in Figure 2.2.

The modulation scheme "Coded Orthogonal Frequency Division Multiplexing" (COFDM) provides a solution to the problems caused by multipath propagation effects (for a detailed discussion see e.g., [She95] or [Sto98]). COFDM is a multicarrier system. In contrast to analogue transmission, the input signal is digitized and the digital data stream is transmitted by appropriate coding onto so-called symbols of duration T_S. Physically, each symbol is the result of the superposition of a set of carriers whose amplitudes and phases can be adjusted to correspond to digital input data stream. After the time T_S has elapsed new amplitudes and phases are chosen according to the input data thus giving rise to another, different symbol, which is then broadcast for a time T_S again.

The structure of COFDM already indicates that it is a modulation scheme that is applicable to digital signals only. There are, in fact, two levels of discretization involved. First, the input data stream is transmitted in terms of successive discrete symbols of duration T_S. Furthermore, the carriers can be adjusted only to defined discrete states. If M carriers are employed and each of them can take one of $2K$ allowed states then each symbol has a data capacity of $M * K$ bits.

But, when using such a technique how is it possible to overcome problems caused by multipath environments? To understand this, it is once again helpful to consider a simple scenario consisting of two signal contributions only. Depending on the relative temporal delay between the two signals the receiver will experience more or less strong constructive or destructive interference at the point of reception. Figure 2.3 sketches some of the superposition possibilities.

In order to decode the transmitted information properly, the receiver must temporally synchronize to the incoming stream of symbols. Then, a portion of length T_W of the received signal is sampled. By applying a Fast Fourier Transform (FFT) the amplitudes and the phases are calculated. Figure 2.4 depicts the relative position of symbols and evaluation window for the case $T_W = T_S$.

In case, the evaluation window T_W is adjusted to coincide completely with symbols of the first path, then nevertheless, it will inevitably take contributions from other symbols via the second path as well. If the delay between the two signals is smaller than the symbol length T_S, the evaluation window will contain contributions from symbol n and from an adjacent symbol. Once the delay becomes larger than the symbol duration T_S, the second path gives rise to destructive interference only, and, hence, a significant reduction of the reception quality.

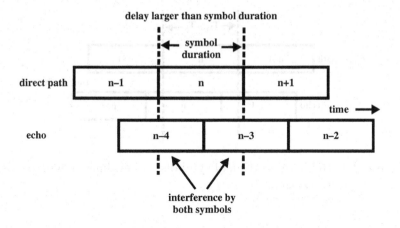

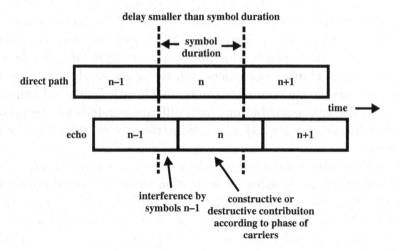

Figure 2.3: Intersymbol interference caused by superposition of two signals that are delayed with respect to each other.

Interference that is caused by a retarded or advanced symbol falling into the evaluation window is called intersymbol interference or self-interference. In a situation, where the time delay is smaller than the symbol duration only a fraction of an unwanted symbol falls into the evaluation window. In this case, there is also a part of the wanted symbol arriving across the second path. Depending on the relative phases of the individual carriers of the two signals, it might lead to constructive or destructive interference.

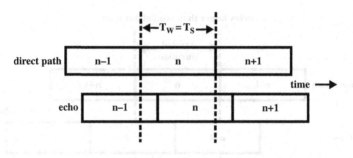

Figure 2.4: Relative position of symbols and evaluation window in the case of $T_W = T_S$.

A straightforward way to overcome the problem of symbols contributing as a total to self-interference in the evaluation window is to introduce a symbol length, which is larger than the utmost in practice appearing time delay. Unfortunately, increasing the symbol duration reduces the data capacity. In principle, this loss can be compensated for by employing a larger number of carriers. On the other hand, as long as the evaluation window T_W is equal to the symbol length T_S like in Figure 2.4, it is inevitable that contributions from different symbols will be present in the evaluation window. Hence, self-interference would always be encountered.

The latter observation triggers the crucial move to avoid even these remaining problems by using an evaluation window whose duration T_W is smaller than the symbol length T_S. The difference $T_G = T_S - T_W$ is called the guard interval. Its length needs to be properly chosen. As indicated above already, choosing a duration of the guard interval that exceeds the largest time delay to be confronted with in a practical situation would resolve the situation. But as pointed out already earlier, this would unacceptably decrease the data capacity of the system. Therefore, a typical value for T_G, which is considered a good balance between protection and reduction of data capacity is, for example, $T_W/4$.

The introduction of a guard interval not only allows to compensate for harmful interference in multipath environments. Rather, this feature of the COFDM system design can be actively employed when setting up transmitter networks. As long as the delay spread at a given point of reception is smaller then the guard interval several signal contributions constructively add. Consequently, a COFDM network can employ several transmitters using the same frequency to broadcast the same content as long as the distance between the transmitter sites is properly chosen. This radical new concept in comparison to analogue networks is called single frequency network (SFN) operation.

As already mentioned, the demodulation of the received symbols is accomplished by utilizing the FFT. A successful application of that technique presumes that the duration of the evaluation window T_W and the spectral separation between two carriers are mutually linked. Hence, it is reasonable to use carriers that are equally distributed across the bandwidth with a spectral distance of

$$\Delta f = \frac{1}{T_W} \tag{2.4}$$

separating them.

In mathematical terms such a set of carriers can be described as an orthogonal system. If the k-th carrier in baseband is written as

$$\Psi_k(t) = e^{ik\omega_W t} \tag{2.5}$$

with

$$\omega_W = \frac{2\pi}{T_W} \tag{2.6}$$

the orthogonality relation

$$\frac{1}{T_W} \int_{t}^{t+T_W} dt' \, \Psi_k(t')\Psi_l^*(t') = \delta_{k,l} \tag{2.7}$$

between two carriers holds. The asterisk indicates complex conjugation while the parameter $\delta_{k,l}$ is the Kronecker symbol

$$\delta_{k,l} = \begin{cases} 1, & k = l \\ 0, & k \neq l \end{cases} . \tag{2.8}$$

Figure 2.5 outlines the position of the individual carriers. Usually an even number of carriers is employed. Therefore, the middle carrier f_C is not used in general.

For the demodulation the received signal is downshifted from RF frequency to baseband. Different kinds of filtering are applied during this process. Then the signal is digitized on the basis of an appropriate

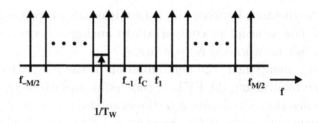

Figure 2.5: Schematic representation of the carriers of a COFDM
signal consisting of M carriers.

sampling rate. From a set of samples corresponding to a time interval
of T_W the Fourier coefficients are calculated by means of the FFT. The
amplitudes and phases of the carriers contain the transmitted information. It has to be noted that the restriction of the evaluation to a finite
time interval implies that the calculated spectrum corresponds to a con-
volution between the COFDM signal and the applied weighting function
used for the FFT. The theoretical representation of Figure 2.5 is replaced
by a spectrum of the form given in Figure 2.6. The underlying weighting
function in that case is a simple rectangular window.

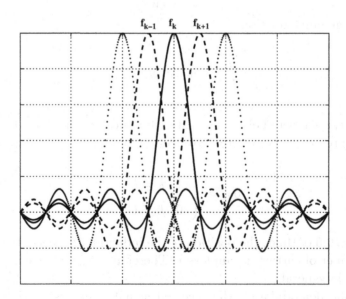

Figure 2.6: Spectrum of a COFDM signal convoluted with a rectangular window.

The spreading of the spectrum does not cause any problems as long as the receiver frequency tuning is correct. The proper choice of the COFDM parameters like width of the evaluation window T_W and the spectral carrier separation guarantee that at the location of the absolute maximum of one particular carrier all other carriers have their nulls. The evaluation of the signals on the basis of FFT is thus not effected.

Clearly, this is no longer true, if self-interference comes into play. Then orthogonality is violated. Taken seriously, this holds for mobile reception, too. In that case, Doppler shifts according to

$$\Delta f^{\text{Doppler}} = f \frac{v}{c} \tag{2.9}$$

appear. The magnitude of the shift depends on the velocity component parallel to the direction of motion of the incoming wave and on the absolute value of the frequency. The quantity c represents the velocity of light. These Doppler shifts are the reason why the spectral location of the carriers and the position of the Fourier coefficient do no longer coincide. As has become clear by now, this is, however, a prerequisite for a successful demodulation.

In principle, COFDM systems are well suited to cope with perturbations caused by Doppler shifts during mobile reception. However, some care has to be taken when defining the COFDM parameters. A good system should work without problems as long as the Doppler shift does not exceed 5% of the intercarrier spacing, that is

$$\frac{\Delta f^{\text{Doppler}}}{\Delta f} < 0.05 . \tag{2.10}$$

With the help of the relation disputed in Equation (2.4) this can be cast into the form

$$\Delta f^{\text{Doppler}} \times T_W < 0.05 . \tag{2.11}$$

Equation (2.11) allows a rough estimate up to which velocity Doppler shifts can be compensated for.

The utilization of a large number of carriers and the introduction of a guard interval are the foundations of any COFDM system. Despite these features, it cannot be avoided that depending on the transmission characteristics the decoding of the information is in general erroneous. If the amplitude of a certain carrier becomes too small due to destructive interference, it is no longer possible to determine its phase correctly with a high probability. The bits derived from a carrier with a small amplitude, thus tend to be wrong then. In case, this problem affects bits that are important, for example, because they carry information about the structure of the entire broadcast signal, the reception quality will degrade dramatically.

A very natural solution to this problem is to add some degree of redundancy to the original bit stream. Then, if a certain amount of bits cannot be properly decoded this will cause no harm to the reception. The receiver will still get all the information that is necessary to decode the signal completely. However, such a robustness of the system has to be paid for in terms of reduction of net data capacity.

Further protection mechanism can be applied as well. Time and frequency interleaving are very common approaches to make the transmission less susceptible to the impact of harmful propagation conditions. Interleaving means to separate adjacent "quantities" from each other according to a defined scheme. The term "quantities" refers to both temporal as well as spectral characteristics of a given signal. For example, the temporal sequence of bits can be reorganized to form a new time sequence that is transmitted instead of the original one. Figure 2.7 illustrates the approach.

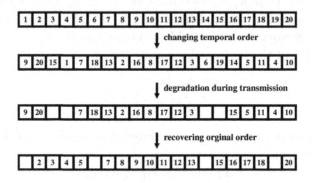

Figure 2.7: Schematic representation of time interleaving.

If due to temporal burst-like perturbations of the transmitted signal a certain time interval is affected, this does not necessarily pose a problem. Two temporarily adjacent bits in the transmitted data stream might be located far enough from each other so that they give rise only to nonharmful degradation for the original signal. The same strategy can be applied to the spectral representation of a signal. To this end, after the application of the FFT carriers are interchanged in order to separate spectrally adjacent carriers. This allows to decrease the harmful impact multipath propagation conditions can have as shown in Figure 2.2. In summary, time and frequency interleaving allow to reduce the influence of localized perturbations both in the time and in the frequency domain. More details concerning interleaving issues can be found, for example, in [Lau96] or [Sto98].

2.2 Digital Audio Broadcasting

The digital terrestrial broadcasting system known as Digital Audio Broadcast (T-DAB) has been drafted around 1990 in the framework of the European research and development program Eureka 147 [EUR96]. It has been standardized in 1997 [ETS97a] and rests on four basic pillars, namely an appropriate source coding technology, special channel coding algorithms, multiplexing of several programs, and COFDM as the modulation scheme for the signal transmission. Figure 2.8 gives an schematic overview of the T-DAB signal generation procedure.

The total available data capacity of the system can be distributed onto several independent data source streams. Both audio and other arbitrary data are allowed. The combination of audio program and data service(s) represents one program content. Each source data stream can have its own data rate. The data rate of the source data is reduced with the help of the data compression method MPEG-1, layer II [ISO93] also known as MUSICAM. Data reduction is achieved by employing psychoacoustic effects [Zwi06]. Basically, this means that only that data is kept that can actually be perceived by the human auditory system. First of all, only frequencies with an amplitude above a frequency-dependent threshold are perceived by the human ear. Furthermore, if, for example, a signal contains two frequencies that are separated only by a few Hertz and at the same time have very different amplitudes, then the human auditory system is not capable to identify, that is to hear the weaker frequency. This effect is called masking. But, consequently, this means that there is no need to transmit this not perceivable information.

Next, protection against propagation distortions of the signal needs to be built into the system. There are three mechanisms that are employed in T-DAB. First, special punctured convolutional codes are applied [Pro89]. The main objective of the application of these kind of

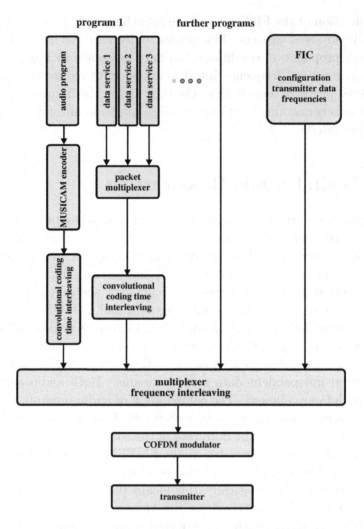

Figure 2.8: Procedure for the generation of a T-DAB multiplex.
FIC stands for "Fast Information Channel".

codes is to add redundant information to the data stream. The amount
of redundant information is quantified in terms of the so-called code
rate. A code rate of 1/2 means that the number of bits in the original
data stream is doubled. Therefore, in principle the full information can
be restored by the receiver if due to the propagation influence half of
the bits are not correctly transmitted. To further protect the transmis-

sion against perturbations, time interleaving as described in the previous section is used for the generation of the T-DAB signal, too.

At the end, all data streams are combined to build a data multiplex. If one audio program is to be supplemented by more than one data service, the corresponding services are bundled into a proper data service multiplex. Channel coding is applied to this data multiplex, too (see e.g., [Lau96]). At the stage of multiplexing all programs together, further protection against propagation degradation is added by using frequency interleaving. The information how the T-DAB multiplex is structured is prepared independently from the program branches and is contained in the so-called Fast Information Channel (FIC).

Together with all program data the FIC is fed into the T-DAB multiplexer whose output goes to a COFDM modulator. Its task is to generate a corresponding baseband signal. This process rests primarily on the application of the FFT. Subsequently, a D/A conversion is carried out and the signal is shifted to the required radio frequency. Clearly, this generation is accompanied by appropriate filtering and amplifying.

The COFDM process utilizes a nominal bandwidth of 1.75 MHz for the generation of the T-DAB signals. Since T-DAB has been designed for mobile reception a very robust modulation scheme was mandatory. Therefore, the differential modulation DQPSK[1] has been chosen. Each carrier of the COFDM signal is assigned one of four allowed different phases. The information that is to be transmitted is coded in terms of phase differences of identical carriers of successive T-DAB symbols.

Each T-DAB signal is built by a sequence of successive COFDM symbols. A number of 76 symbols is grouped to build a so-called T-DAB frame, which is preceded by the null symbol that simply means that during the duration of the null symbol there is no power output of the transmitter at all. This is employed in order to establish a first rough synchronization of the receiver. The null symbol is followed by the phase reference symbol, whose carrier phases are known to the receiver. This constitutes a repeated starting point for the calculation of the phase differences of the carriers of successive symbols. The temporal structure of the T-DAB signal is in line with the parameters of the MUSICAM source coding so that a very quick resynchronisation of the receiver is possible once this is lost due to reception conditions.

---

[1]DQPSK means **D**ifferential **Q**uadrature **P**hase **S**hift **K**eying.

As already discussed in the earlier section, it is possible to adjust the COFDM parameters to adapt the broadcasting system to different coverage environments and targets. For T-DAB, in total four different operation modes have been defined in [ETS97a], which can be utilized under different conditions. Table 2.1 gives an outline of the four sets of allowed COFDM parameters.

Table 2.1: The four possible operation modes for T-DAB.

	Mode I	Mode II	Mode III	Mode IV
Number of carriers	1536	384	192	768
Carriers spacing Δf (kHz)	1	4	8	2
Symbol length T_S (μs)	1246.0	311.5	155.75	623.0
Guard interval T_G (μs)	246.0	61.5	30.75	123.0

Mode I assumes a transmission in the VHF range that in particular is suitable for the coverage of wider areas due to the corresponding wave propagation characteristics of VHF. In order to reduce degradations caused by violations of the guard intervals the network implementation should take into account an inter-transmitter distance that should not exceed 73 km significantly. This distance corresponds exactly to the lap electromagnetic waves can travel within the period of a guard interval of 246.0 μs.

Mode II has its focus on L-Band application. According to the smaller guard interval the inter-transmitter distance in single frequency networks should be reduced as well. Therefore, only smaller areas can be covered without problems. This goes hand-in-hand with the impaired wave propagation conditions in comparison with the VHF range. Nevertheless, for large city coverage this mode is very well suited.

The third mode has originally been defined for satellite broadcasting. In that case, large Doppler shifts due to the high velocities of the satellites relative to the ground have to be accounted for. Therefore a carrier spacing of 8 kHz has been chosen. In contrast to terrestrial broadcasting, echoes by reflections do not play a significant role for satellite distribution. Hence, the relatively small guard interval is justified. Finally, a fourth mode has been defined for coverage tasks that lie in between the wide area and the city only case.

The cycle to pass through during reception in order to detect and provide the requested audio signal or data service is sketched in Figure 2.9. The T-DAB signal is received by the antenna and passed through an adequate process of filtering, downshifting from RF frequency to baseband and digitizing. The COFDM demodulator retrieves the phases and amplitudes of the used carriers. This is accomplished by applying an inverse FFT. The FIC is analyzed that allows to restrict the data reconstruction to those parts of the entire bit stream that actually contain the requested audio program or data services. Errors that were caused by the transmission can be removed by making use of error correction algorithms unless the damage of the signal is not irreparable. Usually, soft-decision Viterbi methods are utilized for those purposes. They represent enhancements of the original Viterbi algorithm [Vit67]. The resulting bit stream is then passed to a playback unit or an adapted data output device.

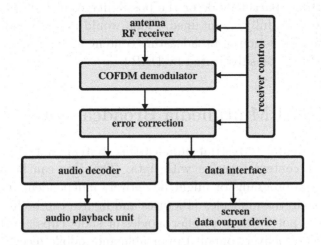

Figure 2.9: General demodulation procedure for the reception of a T-DAB signal.

2.3 DAB+

The source coding employed for T-DAB has been under criticism for a long time. T-DAB has been standardized more than ten years ago and it was argued that MPEG-1, layer II is outdated. More efficient coding schemes have been introduced in the meantime. In 2007, an enhancement of T-DAB that is called DAB+ has been standardized [ETS07]. There are two significant changes in comparison to T-DAB. First, MUSICAM has been substituted by the more advanced coding scheme HE AAC v2 [ISO05a]. An overview about this coding scheme can be found, for example, in [Mel06]. This allows higher data reduction rates compared to T-DAB. Second, the channel coding has been enhanced as well by adding Reed-Solomon coding to make the transmission more robust [Ree60].

The audio program is transmitted as a standard T-DAB data stream after having been encoded by HE AAC v2 and Reed-Solomon. This is actually the reason why in principle it is possible to combine DAB and DAB+ content. In other words, a DAB+ multiplex can be build from audio programs coded with MPEG-1, layer II, that is standard T-DAB content, together with others that employ HE AAC v2 [Wor07]. The system is very flexible in that respect. DAB+ has been optimized to carry audio content rather then video. This is reflected in the fact that video codecs are not supported. Furthermore, all features incorporated into T-DAB like packet mode or the possibility to include program associated data are fully maintained. This would allow broadcasters to continue the production of broadcasting content without any change, in case T-DAB is to be substituted by DAB+.

2.4 Digital Multimedia Broadcasting

T-DAB is a digital terrestrial broadcasting system that allows to distribute audio content together with data. The data can be associated with the audio programs or might be entirely independent from any of the programs in the multiplex. Pictures and figures can be broadcast as well, but it is not foreseen to offer movies or short clips. Furthermore, T-DAB aims to deliver content to portable and mobile receivers.

In particular the latter feature has become more and more important in recent years. Customers are keen to have access to audio and video programs while on the road or in trains. Consequently, the idea emerged to build a system that would allow to satisfy exactly this demand. Digital Multimedia Broadcasting (DMB) as an extension of T-DAB has been developed for that purpose.

For the transmission of audio and video programs to portable and mobile receivers most likely being equipped only with rather small screen and limited storage capacity, a mixture of different source coding schemes has been employed. Audio content is compressed on the basis of MPEG-4/AVC [ISO05b], while audio programs are encoded with the help of HE AAC v2 [ISO05a] as in the case of DAB+. In contrast to DAB+, DMB has been optimized to broadcast television content. Even though in principle, audio programs can be broadcast as well via DMB, this is usually not recommended. A detailed comparison between DAB+ and DMB can be found at the Website of the WorldDMB Forum [Wor07].

2.5 Digital Video Broadcasting

At the beginning of the 1990s of the last century digital terrestrial broadcasting (DVB-T) has been developed. Several international organizations like the European Telecommunications Standards Institute (ETSI) [ETS08], the European Committee for Electrotechnical Standardization (CENELEC) [CEN08] and the European Broadcasting Union (EBU) [EBU08] have been involved. DVB-T constitutes an open standard as does T-DAB.[2]

Television broadcasting has a different focus than audio broadcasting, there are different requirements to be met. In the first place, a significantly larger technical effort than in the case of audio broadcasting is necessary. This is related to the data capacities that are required to provide a satisfying television service. They exceed those of typical audio programs by an order of magnitude. This did not change in the digital age either. Furthermore, the acceptable error rates for television broadcasting are significantly smaller at the same time. Moreover, terrestrial broadcasting as one of the major distribution paths for audio and video content has no longer the same importance for television as this is still the case for radio. There certainly exist variations across Europe, but the majority of people either uses cable or satellite links for primary

---

[2]There are different definitions of the term open standard. However, publication of all details of the standard accessible to anybody free of royalty fees are common elements of all definitions.

television reception. This allow the customers to obtain a very large number of different programs. In the case of satellite, several hundreds of programs can be received in Europe.

Consequently, a successful introduction and market penetration of DVB-T certainly calls for more programs that can be provided via terrestrial distribution. Programs, so far exclusively distributed via cable or satellite should be broadcast terrestrially, too. This alone would probably not be sufficient. But, DVB-T has more to offer, since it allows to provide television services for portable and even mobile reception, as well. Both cable and satellite reception do not offer these possibilities. Customers have to attach their television set to either a satellite dish or a cable network, both of which by definition are fixed to a certain location.

Analogue television has been planned for fixed roof-top reception. Even though portable analogue reception might be feasible in the vicinity of a transmitter, mobile reception does not usually work. But, in a globalized world where people are constantly on the move fixed antenna reception of audio and television programs can no longer be a general coverage target. Portable and mobile reception have become more important in recent years due to a significant change of costumer's daily life circumstances and habits. These two reception modes offer a clear added value under these circumstances. From a customer's point of view it must be possible to carry the DVB-T receiver outside to be used during different spare time activities. This implies that DVB-T terrestrial reception has to be independent of a complicated roof antenna or satellite dish connections. Instead, a simple-pole antenna must suffice.

As in the case of T-DAB several television programs are bundled to form a multiplex. Figure 2.10 sketches the generation of a DVB-T signal.

A program consists of video signals, audio signals, and pure data. All the three types of data undergo data reduction procedures based on MPEG-2. MPEG-2 offers the freedom to assign different data rates to each of the programs in the multiplex in an independent manner. This means the data rate for each of the programs can be adapted to comply with predefined coverage targets. Once the data reduction is accomplished for the video, audio, and data part of a single television program, they are bundled into a submultiplex. Together with further programs and service information the DVB-T multiplex is subsequently built.

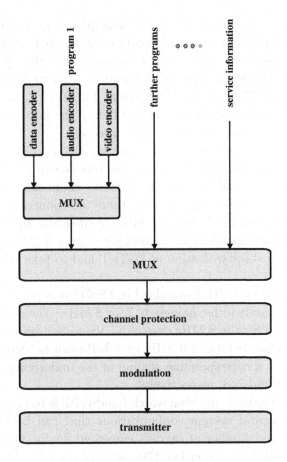

Figure 2.10: Generation of a DVB-T multiplex.

Channel coding is the next step. Several mechanisms are applied for that purpose. Reed-Solomon and punctured convolutional codes are employed in order to make the transmitted data more rugged against transmission errors. In order to further enhance the Reed-Solomon coding a bit interleaving step is introduced before the convolutional coding is carried out. Finally, the baseband signal is generated by a COFDM modulator. Details of the whole process can be found in [Rei01] or in the standard of DVB-T itself [ETS97b].

Comparing Figure 2.8 and 2.10 a significant difference between T-DAB and DVB-T can be detected. In the case of T-DAB data reduction and channel coding are applied before the multiplex is generated, while for DVB-T the protection against propagation influences is added only after the multiplex has been built. As a consequence, in the case of T-DAB there is no need to decode the entire signal in order to access a program chosen by the customer. This is different for DVB-T, where first the entire data stream needs to be decoded before the information relating to a particular program can be further processed.

In Europe, there are several channel rasters used in the spectrum ranges decided by the ITU for television broadcasting. Basically, the spectrum bands are subdivided into 8 MHz channels. This holds in particular for the UHF range. In VHF there are countries, in particular European countries, which use a 7 MHz bandwidth. In other parts of the world there are also channel rasters based on a 6 MHz spacing. It is obvious that the standardization of DVB-T had to take these facts into consideration.

According to the DVB-T standard [ETS97b] it is possible to use one of the common bandwidths, namely 6, 7, or 8 MHz. The system has been designed initially for the 8 MHz case only. Values for system parameters connected to a bandwidth of 6 MHz or 7 MHz can be derived from the 8 MHz values by a corresponding scaling of the underlying system clock by a factor 6/8 and 7/8, respectively.

Apart from the basic decision which bandwidth is to be utilized, there are two fundamental system configurations that can be implemented. They differ by the number of carriers employed for the COFDM modulation. It is possible to use either 1705 or 6817 carriers. They are called 2k or 8k mode, respectively. Depending on the used bandwidth, different durations of the evaluation window T_W and the carrier spacing Δf result. Table 2.2 summarizes the most important parameters.

For the two modes, a total of six different guard intervals has been defined. The four values $T_G = 224, 112, 56,$ and $28\,\mu s$ can be used in 8k mode while for the 2k mode $T_G = 56, 28, 14,$ and $7\,\mu s$ are allowed. So, the values $56\,\mu s$ and $28\,\mu s$ are available for both modes. If two transmitters in a DVB-T single frequency network are separated by more than a distance $\Delta r = c \times T_G$ self-interference can result. The quantity c denotes the velocity of light. As a consequence, the choice of the mode and the value of the guard interval is not a totally free one. It is to a large extent determined by the coverage target. For wide-area coverage across an entire national territory in terms of an SFN structure there is, however, not much freedom left. The only promising DVB-T variant thereto should rest on the utilization of the 8k mode together with $T_G = 224\,\mu s$.

Table 2.2: Potential COFDM parameters for DVB-T [ETS97b].

	2k mode			8k mode		
Bandwidth of TV channel [MHz]	6	7	8	6	7	8
Evaluation window T_W [μs]	298	256	224	1194	1024	896
Carrier spacing Δf [Hz]	3348	3906	4464	837	977	1116
DVB-T Bandwidth B [MHz]	5.72	6.66	7.612	5.71	6.66	7.609

In contrast to T-DAB, it is possible to employ several different modulation schemes. Either QPSK, 16-QAM or 64-QAM can be applied.[3] The amount of data that can be transmitted increases along this ordering of modulations. On the other hand, the transmission becomes less rugged at the same time. In fact, from QPSK to 64-QAM an increasing protection ratio between the useful and the unwanted signal contributions has to be taken into account. As a matter of fact, more elaborate network structures might be needed. Fortunately, DVB-T offers the possibility to adjust different error protection levels. This can be used to counterbalance the consequences of higher modulation schemes.

DVB-T does not use a differential modulation scheme. Therefore it is necessary to dedicate a fraction of the total data capacity for synchronization purposes. A subset of the total number is utilized as pilot carriers. They have precisely defined amplitudes and phases that are known to the receiver. There exist two types of pilots. The first type has fixed positions within the used bandwidth. Furthermore, there are pilots that change their position within the spectrum from one symbol to the next. The way they move is purely deterministic and also known to the receiver. This offers additional protection for the synchronization against degradation caused by narrow band fading as a consequence of multi-path propagation conditions.

---

[3]QPSK means **Q**uadrature **P**hase **S**hift **K**eying whereas QAM stands for **Q**uadrature **A**mplitude **M**odulation.

The net data rate of DVB-T is independent of the chosen mode. Both 2k as well as 8k allow the transmission of the same amount of data per second. It is true that the 8k mode employs four times more carriers than in the 2k case. But, at the same time the symbol length is four times as large for 8k variants, so that after all the data capacity remains the same. The crucial factors determining the data capacity are the modulation scheme applied, the error protection level, the duration of the guard interval, and the bandwidth used. By varying these parameters a huge variety of different operational system variants can be put into practice. Table 2.3 presents the most important possibilities. For a more profound discussion it is referred to [Rei01].

In the case of T-DAB as well as for DVB-T other planning parameters such as minimum required field strength or protection ratio between wanted and unwanted signal contributions have to be taken into account as well. A full presentation of all these quantities would lie outside of the scope of this book. Very detailed tables containing all this information have been published both by EBU and ITU (see e.g., [ITU00], [EBU98], [EBU01], [EBU02]). They are constantly updated. The Final Acts of the Regional Radiocommunication Conference [ITU06] held in 2006 in Geneva that established a frequency plan for T-DAB and DVB-T, provides the most recent update of all relevant planning parameters.

The total data capacity a certain combination between modulation and code rate provides is usually employed to generate a DVB-T multiplex containing 4–6 television programs in PAL quality. However, in principle it is also possible to utilize the available capacity to broadcast 1–2 programs in HDTV quality. Detailed in formation on HDTV can be found, for example, in [Woo04]. Even though from a technical point of view this is certainly feasible it has to be borne in mind that HDTV via DVB-T will result in a demand for frequencies similar to that of analogue television. Therefore, a decision between more programs or few programs with significantly better quality would need to be taken by broadcasters, network operators, and regulators.

2.6   DVB-T2

Similar to the situation of T-DAB also for DVB-T there has been a discussion about the source coding technology MPEG-2. The DVB-T standard has been issued in 1997, and, hence, the employed source coding

Table 2.3: Net data rates for different DVB-T operation modes in the case of an 8 MHz TV channel [ETS97b].

Modulation	Error protection ratio	Net bit rate (MB/s)			
		$\Delta/T_W = 1/4$	$\Delta/T_W = 1/8$	$\Delta/T_W = 1/16$	$\Delta/T_W = 1/32$
QPSK	1/2	4.98	5.53	5.85	6.03
QPSK	2/3	6.64	7.37	7.81	8.04
QPSK	3/4	7.46	8.29	8.78	9.05
QPSK	5/6	8.29	9.22	9.76	10.05
QPSK	7/8	8.71	9.68	10.25	10.56
16QAM	1/2	9.95	11.06	11.71	12.06
16QAM	2/3	13.27	14.75	15.61	16.09
16QAM	3/4	14.93	16.59	17.56	18.10
16QAM	5/6	16.59	18.43	19.52	20.11
16QAM	7/8	17.42	19.35	20.49	21.11
64QAM	1/2	14.93	16.59	17.56	18.10
64QAM	2/3	19.91	22.12	23.42	24.13
64QAM	3/4	22.39	24.88	26.35	27.14
64QAM	5/6	24.88	27.65	29.27	30.16
64QAM	7/8	26.13	29.03	30.74	31.67

can no longer be considered state-of-the-art. Other algorithms have been developed in the meantime, in particular MPEG-4 is currently the most efficient machinery to prepare audio and video data for distribution via terrestrial broadcasting systems. Consequently, DVB-T has been extended to make use of MPEG-4 to achieve higher data reduction rates. This can be used to provide more television content in the first place.

However, the amount of data rate that can be released can also be utilized to increase the amount of redundant information in the digital signal, and therefore, leads to more robustness against propagation influences. In relation to a potential distribution of HDTV via terrestrial networks the application of DVB-T2 instead of DVB-T might be a significant advantage in terms of efficient usage of the available spectrum.

The standardization of DVB-T2 is expected to be finalized by the end of 2008. Shortly after a DVB-T2 prototype chip will be available and promoters of DVB-T2 are planning for a mass-market launch in 2009/2010. Since it is very likely that DVB-T will have reached a deep-market penetration it is evident that DVB-T2 needs to be put into operation in a manner that is as compatible as possible with the then existing standard DVB-T environment. In a first step, fixed reception by domestic DVB-T antenna systems is envisaged. Transmissions shall be designed such that compatibility with DVB-T broadcasting can be assured. More information can be found at the Website of the DVB Project [DVB08].

2.7   Digital Video Broadcasting – Handheld

Portable and mobile reception is becoming a more and more important issue, both for network providers and for providers of any kind of telecommunication services, including broadcasters. This led to the demand that television services should also be receivable under these conditions. DVB-T as it has been standardized in [ETS97b] is not the appropriate system. Under certain conditions, for example, using antenna diversity in order to boost the antenna gain it is possible to achieve mobile reception for DVB-T. But effectively the required effort to do so is not acceptable.

Therefore, a variant of DVB-T has been designed that should be able to provide broadcasting services in particular for portable and mobile usage with acceptable quality. This means that in the first place a handheld receiver has to be targeted at. This includes multimedia mobile phones with color displays, as well as personal digital assistants, or pocket PC types of receivers. All these devices have one thing in common, namely that they are rather small, having only light weight and— very important—are energized by batteries. Apart from that, portable and mobile reception naturally includes indoor reception, sometimes even deep indoor, that is in basements or deep inside concrete buildings. This requirement is, however, in conflict with the small dimensions of the receiving devices, since handheld devices employ built-in antennas. These usually have rather poor receiving characteristics both in terms of antenna gain as well as directivity. A multiantenna diversity approach to improve the receiving characteristics is all but impossible under such conditions.

In November 2004, the DVB-H standard has been published by ETSI [ETS04a]. DVB-H is to large extent compatible with DVB-T. This has been explicitly taken care of when designing the system because one of the requirements in particular of broadcasters was to be able to implement DVB-H networks with the help of existing DVB-T networks, too.

Nevertheless, several major changes in relation to DVB-T have been introduced. The energy problem linked to the battery operation of receiving devices has been tackled by introducing a special power-saving mechanism called time slicing. In the case of DVB-T, the whole data stream has to be decoded before individual programs can be accessed. This poses a severe problem for handheld devices powered by batteries due to high-power consumption. For the DVB-H standard this problem has been resolved by transmitting the data associated with a particular service not continuously but only throughout dedicated time slices. In between these slices, when other DVB-H services are broadcast the receiver switches to a power-saving mode [ETS04b].

Furthermore, an enhanced error-protection scheme has been incorporated. It is called "Multi-Protocol-Encapsulation — Forward Error Correction" (MPE-FEC). A prerequisite of this is that in contrast to DVB-T where the DVB transport stream is based on MPEG-2, DVB-H is based on IP. This is accomplished by adapting the DVB Data Broadcast Specification to allow for the "Multi-Protocol–Encapsulation" [ETS04b]. On the level of MPE additional forward error protection is added which is MPE-FEC. It basically consists of a special Reed-Solomon code together with a block interleaver. MPE-FEC imposes a frame structure that is aligned with the time slicing technology of DVH-H. Figure 2.11 schematically shows the generation of a combined DVB-T/DVB-H signal. More details can be found, for example, in [Kor05] or [Far06].

A further modification of the DVB-T standard for DVB-H relates to the incorporation of an additional COFDM mode. DVB-T allows to use either the 2 k or the 8 k mode. DVB-H can be operated in terms of a 4k mode as well. Table 2.4 shows the differences of the three modes for some COFDM parameters for the case of a 8MHz channel.

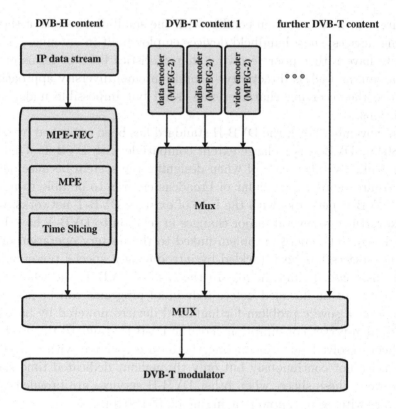

Figure 2.11: Generation of a DVB-T/DVB-H multiplex.

The 4k mode has been introduced in order to allow for network structures that can benefit from both DVB-T modes, namely 2k and 8k. Due to larger guard intervals in comparison to 2k mode, 4k-DVB-H operated in SFN mode allows for a less denser network, that is the intertransmitter distance can be increased without causing self-interference. Moreover, the susceptibility to Doppler shift is reduced compared to the 8 k mode. This is particularly important in relation to providing services for mobile reception.

2.8   MediaFlo

All digital terrestrial broadcasting systems described so far have been developed by international organizations or consortia and constitute open standards. In contrast, MediaFlo is a proprietary digital terrestrial

Table 2.4: COFDM parameters for DVB-H.

	2k mode	**4k mode**	**8k mode**
Number of carriers	1705	3409	6817
Evaluation window T_W [μs]	224	448	896
Guard intervals T_G [μs]	7,14,28,56	14,28,56,112	28,56,112,224
Carrier spacing Δf [Hz]	4464	2232	1116

broadcasting system even though the standardization with ETSI has been initiated. Nevertheless, it is interesting, because it has been developed under different conditions than T-DAB or DVB-T. MediaFlo was designed from the very beginning to be a broadcasting system that is to be combined with mobile communication. Hence, one of the assumptions was to employ existing cellular network structures rather than typical broadcasting networks. Furthermore, no backward compatibility constraints had to be taken into account as, for example, in the case of DVB-H.

The design of the system therefore focused on the task to deliver multimedia content to handheld devices, in particular mobile phones in an optimal manner. As in the case of DVB-H this imposes the constraint that power consumption needs to be properly taken into account. Therefore, an efficient power-saving technique needed to be applied.

In general terms, the overall system design looks very similar to DVB-H. MediaFlo incorporates advanced forward error correction techniques involving the concatenation of a parallel concatenated convolutional code, that is a Turbo coding technology, and a Reed-Solomon code. The modulation scheme is COFDM where a 4k mode is employed. Each carrier can be coded either by QPSK or 16 QAM. Four signals bandwidths are supported, namely 5, 6, 7, and 8 MHz. In contrast, to DVB-H the modulation has a layered hierarchy to support transmissions of base and enhancement layers with different levels of robustness. As for any other nondifferential modulation scheme, part of the COFDM carriers are employed as pilot carriers, giving the receiver the possibility to synchronize the signal.

Furthermore, MediaFlo supports fine-grained multiplexing that allows the implementation of a variable bit rate encoding. The available bandwidth can be dynamically allocated on a per second basis. In contrast, DVB-H allocates a fixed bandwidth per time slice. One time slice is the smallest unit of access to a DVB-H multiplex. The DVB-H receiver must decode the entire time slice no matter how small the data for the wanted service contained therein.

By design, DVB-H and DVB-T content can coexist within a single DVB-T multiplex. This is not foreseen with MediaFlo. Rather, an independent network is necessary. Further details concerning MediaFlo can be found on the QUALCOMM Website [Qua07] while the physical layer of MediaFlo is described in detail in [Cha07].

2.9   Digital Radio Mondiale

Terrestrial broadcasting can take place in several different frequency ranges. The digital terrestrial broadcasting systems described in the preceding sections are intended to be put into operation in the frequency bands III, IV, and V, and L-Band, that is 174–230 MHz, 470–606 MHz, 606–862 MHz and 1452–1479.5 MHz. All of them are considered as broadband systems, because they occupy between 1.75 and 8 MHz.

In addition to these system there exists another interesting digital terrestrial broadcasting to be mentioned here briefly even though it is not meant for operation in one of these bands. It is called Digital Radio Mondiale (DRM), which is a COFDM system, too. According to the standard [ETS05] DRM can be used below 30 MHz, that is in the short- and medium-wave regime. In the first place, DRM is a digital terrestrial broadcasting system that is to substitute the analogue AM transmissions. Therefore, it employs a bandwidth of 9 kHz or 10 kHz only. Compared to the systems discussed in the preceding sections, this is very narrowband. However, such a bandwidth has been chosen to fit DRM into the existing AM channel raster.

Based on the coding scheme MPEG-4 HE AAC v2 [ISO05a] a bandwidth of 10 kHz allows to obtain between 8 kB/s and 20 kB/s depending on the amount of data capacity that needs to be dedicated to achieve a certain degree of ruggedness against propagation perturbations. HE

AAC v2 is the right choice for audio content. For speech programs, other coding schemes such as MPEG-4 CELP or MPEG-4 HVXC (which are both part of the MPEG-4 family) can be utilized. These coding schemes are particularly adapted to these kind of input signals.

Similar to all other systems described here the program input streams are bundled into one multiplex. Channel coding to protect the transmission against propagation errors is added, too, as well as time interleaving. In order to allow the receiver to synchronize to the signal pilot carriers are included as well. A very detailed description of the DRM system can be found on the Website of the DRM forum [DRM07].

It comes as no surprise that for the remaining part of the spectrum that is also used for broadcasting, namely bands I and II (45–85 MHz and 87.5–108 MHz), also digital terrestrial broadcasting systems are discussed to substitute analogue transmission like FM radio in the future. Therefore, within the DRM consortium a process has been initiated to extend the DRM such that this system could be used in band I and II as well. The name DRM+ or DRM120 has been coined for this extension. Similar to the situation below 30 MHz a bandwidth is proposed that would fit into existing channel rasters. For band II this means a bandwidth of 100 kHz to fit into the FM structure as defined for Europe. The DRM standard will be extended to cover this kind of digital terrestrial broadcasting system as well.

Chapter 3

Management of the Electromagnetic Spectrum

Telecommunication is omnipresent today. Nearly every household in the Western world owns several radios and most of the families have at least one or sometimes even more television sets. In some countries, for example in Scandinavia, in already 2003 the number of customers having exclusively a mobile phone contract exceeded the number of cable based subscriptions. More wireless telecommunication systems are pushing into the market trying to gain ground and customers.

All these systems are utilizing a part of the electromagnetic spectrum. To guarantee interference free coexistence several international organizations are monitoring and controlling the spectrum usage. In the first place, this task is fulfilled by the International Telecommunications Union (ITU) and with focus on Europe by the Conference Européenne des Administration des Postes et des Télécommunications (CEPT). In recent years the European Commission (EC) started to get more and more involved into the field of international frequency management in order to harmonize the spectrum usage amongst members of the European Union (EU).

3.1 International Telecommunications Union

The ITU [ITU07] has been established 135 years ago. Nowadays, it constitutes one branch of the United Nations (UN) and is subject to UN's rules of procedure. It is the basis for successfully coordinating the usage of the electromagnetic spectrum on a global level. The ITU was created to act as an impartial international organization giving a framework in which national governments represented by their administrations and industries can work together in order to operate telecommunication networks and provide services. Moreover, the further development of telecommunication technologies is an important issue dealt with by the ITU as well.

As a matter of course, everyday people around the globe use their telephones to talk to each other. In the past, this came to pass mainly via fixed telephone connections, in the meantime mobile phones are getting more and more important. Access to the Internet, sending and receiving an e-mail has become irreplaceable both in the business sector and in private activities. Travelling is more and more dependent on telecommunication services. This refers to planning business or leisure trips via Internet or relying on navigation systems based on GPS when being on the road. Short range devices are penetrating our daily lives, which means that any new car sold is equipped with corresponding devices to open and close the doors. All these examples make use of some kind of telecommunication system, which benefits from the work of the ITU of managing the radio-frequency spectrum.

The ITU maintains and tries to extend international cooperation between all its Member States in order to allow for a rational use of any kind of telecommunication systems. Organizations and companies in the field of telecommunication are encouraged to participate in all the corresponding activities, that is research, development, and standardization. One of the main objectives of the ITU is to offer technical assistance to developing countries for mobilizing any kind of resources to improve access to telecommunication services in such countries.

The structure of the ITU reflects its main tasks. It is subdivided into three sectors, namely Radiocommunication (ITU-R), Telecommunication Standardization (ITU-T) and Telecommunication Development (ITU-D). Their activities cover all aspects of telecommunication, from setting standards to improve telecommunication infrastructure in the developing world. Each of the three ITU Sectors works through conferences and meetings, where members negotiate the agreements that serve as the basis for the operation of global telecommunication services. Study groups made up of experts drawn from leading telecommunication organizations worldwide carry out the technical work of the Union, preparing the detailed studies that lead to authoritative ITU recommendations. Figure 3.1 gives an overview.

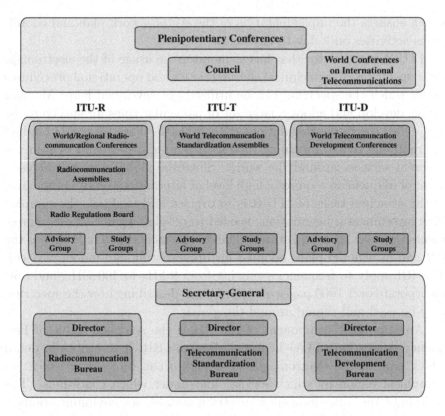

Figure 3.1: Organizational structure of the ITU.

ITU-R draws up the technical characteristics of terrestrial and space-based wireless services and systems, and develops operational procedures. It also undertakes the important technical studies that serve as a basis for the regulatory decisions made at radiocommunication conferences. In ITU-T, experts prepare the technical specifications for telecommunication systems, networks, and services, including their operation, performance, and maintenance. Their work also covers the tariff principles and accounting methods used to provide international service. ITU-D experts focus their work on the preparation of recommendations, opinions, guidelines, handbooks, manuals, and reports, which provide decision-makers in developing countries with "best business practices" relating to a host of issues ranging from development strategies and policies to network management. Each sector also has its own bureau, which ensures the implementation of the sector's work plan and coordinates activities on a day-to-day basis.

ITU-R is the sector that has to monitor the usage of the electromagnetic spectrum. The technical characteristics and operational procedures under which the spectrum can be utilized are developed here. Member States develop and adopt a large set of particular rules for spectrum usage, called the Radio Regulations (RR) [ITU04]. They serve as a binding international treaty governing the use of the radio spectrum by some 40 different services around the world. Since the global use and management of frequencies requires a high level of international cooperation, one of the principal tasks of ITU-R is to oversee and facilitate the complex intergovernmental negotiations needed to develop legally binding agreements between sovereign states. These agreements are embodied in the RR and in regional plans adopted for broadcasting and mobile services. The RR apply to frequencies ranging from 9 kHz to 400 GHz, and now incorporate over 1000 pages of information, describing how the spectrum may be used and shared around the globe.

An important component of the RR is the so-called Table of Frequency Allocation (TFA) in Article 5 of the RR. It describes in detail which part of the electromagnetic spectrum can be used in which geographical region by which service, and under which conditions. The portion of the radio-frequency spectrum suitable for communications is divided into "blocks", the size of which varies according to individual services and their requirements. These blocks are called "frequency bands", and are allocated to services on an exclusive or shared basis. The full list of services and frequency bands allocated in different regions forms the TFA. Even though a particular frequency band might be allocated to a special service like broadcasting, mobile, or fixed service, this can be overruled or extended by means of footnotes containing special arrangements between individual countries.

Changes to the TFA and to the RR themselves, can only be made by a World Radiocommunication Conference (WRC). Alterations are made on the basis of negotiations between national delegations, which work to reconcile demands for greater capacity and new services with the need to protect existing services. If a country or group of countries wishes a frequency band to be used for a purpose other than the one listed in the TFA, changes may be made provided a consensus is obtained from other Member States. In such a case, the change may be indicated by a footnote, or authorized by the application of a RR procedure under which the parties concerned must formally seek the agreement of any other nations affected by the change before any new use of the band can begin.

In addition to managing the TFA, a WRC may also adopt assignment plans or allotment plans for services where transmission and reception are not necessarily restricted to a particular country or territory. In the case of assignment plans, frequencies are allocated on the basis of requirements expressed by each country for each station within a given service, while in the case of allotment plans, each country is allotted frequencies to be used by a given service, which the national authorities then assign to the relevant stations within that service.

With the help of its Bureau ITU-R acts as central registrar of international frequency use, recording and maintaining the Master International Frequency Register (MIFR). It contains entries for more than a million terrestrial frequency assignments and more than 100000 entries relating to different satellite services. Furthermore, ITU-R is the central organization coordinating efforts that ensure that all the different telecommunication services can coexist without causing harmful interference to each other. Several computer based tools are offered to Member States to carry out corresponding analyses.

ITU-R prepares the technical groundwork, which enables radiocommunication conferences to make sound decisions, developing regulatory procedures and examining technical issues, planning parameters, and sharing criteria with other services in order to calculate the risk of harmful interference.

3.2 European Conference of Postal and Telecommunications Administrations

The European Conference of Postal and Telecommunications Administrations (CEPT)[4] [CEP07] is a European regional organization dealing with postal and telecommunications issues and presently has members from 48 countries. It was founded in 1959. Its basic objective is to deepen the relations between members, promote their cooperation, and contribute to the creation of a dynamic market in the field of European posts and electronic communications. Any European country can become a member of CEPT as long as it is a member of the Universal

---

[4]The acronym "CEPT" stems from the French name "Conference Européenne des Administration des Postes et des Télécommunications".

Postal Union (UPU) [UPU07] or a Member State of the ITU. Representatives of ITU and UPU are usually invited to assemblies of CEPT while other intergovernmental organizations may be invited to participate as observers. This is also possible for organizations having signed a memorandum of understanding with CEPT declaring to subscribe to the rules of procedure of CEPT. Finally, the European Commission (EC) and the Secretariat of the European Free Trade Association (EFTA) are invited to participate in CEPT activities in an advisory manner, with the right to speak but not to vote.

The CEPT shows a hierarchic structure of several bodies collaborating in clearly defined way. The highest body of CEPT is the Assembly that is chaired by the Presidency. The latter also acts as the secretariat for the Assembly that adopts major policy and strategic decisions and recommendations within the postal and electronic communications sectors. Committees may be set up by the Assembly dealing with different tasks assigned to them. Currently (2008) there are two committees, namely the European Committee on Postal Regulation/Comité Européen de Réglementation Postale (CERP) and the Electronic Communications Committee (ECC) responsible for radiocommunications and telecommunications.

Each committee has several Working Groups dedicated to special aspects of postal and telecommunications issues. The Committees and the Working Groups are supported by the European Radiocommunications Offices (ERO) located in Copenhagen. ERO is the distribution point for all ECC documentation and also provides detailed information about the work of the ECC and its Working Groups via the ERO Website [ERO07a]. The major responsibilities of ERO includes the drafting of long-term plans for future use of the radio frequency spectrum at a European level. Clearly, national frequency management authorities of CEPT members are supported in their work. Consultations on specific topics or the usage of parts of the frequency spectrum are conducted by ERO. An important task is also the publication of ECC decisions and recommendations and to keep record of the implementation of telecommunication services in Europe. Figure 3.2 gives an overview of the CEPT structure.

From the six Working Groups currently emanating from the ECC there are three having outstanding meaning for the usage of the electromagnetic spectrum in Europe. The first one is Working Group Frequency Management (WGFM), the second is called Working Group Spectrum

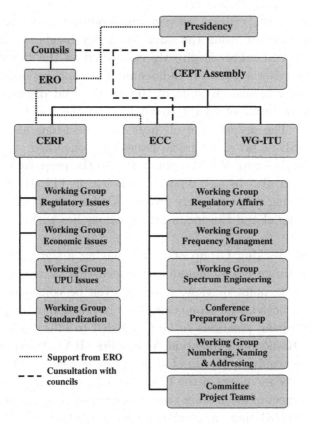

Figure 3.2: Organizational structure of CEPT.

Engineering (WGSE) and finally there is the Conference Preparatory Group (CPG). WGFM is covering all issues that are connected to allocating spectrum to different telecommunication services such as the harmonization of spectrum for Short Range Devices (SRD) or the future of the Terrestrial Flight Telephone System (TFTS) bands. Broadband Fixed Wireless Access in the 3.5 GHz and 5.8 GHz bands are other issues, as well as any activities connected to the harmonization of frequency bands for IMT-2000 (International Mobile Telecommunications 2000) [IMT00]. Broadcasting issues are covered in terms of the preparation of World Radiocommunication Conferences of the ITU.

In WGSE, more technical issues of spectrum usage are dealt with. In particular, this refers to the preparation of technical guidelines for the use of the frequency spectrum by various radiocommunication services.

Furthermore, sharing criteria between radiocommunication services, systems, or applications, using the same frequency bands are studied and assessed by WGSE. This is directly connected to the investigation of compatibility criteria between radiocommunication services using different frequency bands. Results of the studies and discussions are to be published in terms of CEPT-Recommendations and CEPT-Reports as necessary. The preparation of draft ECC decisions also lies in the scope of WGSE as far as technical issues are concerned. As any other CEPT Working Group, WGSE contributes to the preparation of WRCs of the ITU and also to any related work within ITU-R, for example, the corresponding relevant study groups. Collaboration with ETSI and other international and European organizations lies within the WGSE's remit, too.

The third Working Group of CEPT having a significant relevance for spectrum usage issues including broadcasting system is the CPG. From a strategic point of view, it might even be considered the most important one because its scope comprises in particular the preparation of agreed European positions to be forwarded to ITU conferences such as WRCs or Radiocommunication Assemblies (RA). Moreover, CPG is to develop, as required, coordinated positions to assist CEPT Administrations that are members of the ITU Council in presenting a European position in respect of discussions concerning conference agendas and timing. Within ITU-R there also exists a group, dealing with the preparation of ITU conferences, called Conference Preparatory Meeting (CPM) to which CPG contributes as well. European views for WRCs or RAs are typically expressed in terms of so-called European Common Proposals (ECP) whose preparation is supervised by the CPG. In addition, ECP's explanatory documents, called Briefs, are also developed under the control of CPG. These Briefs are intended to support the members of CEPT national delegations in order to present the European positions at WRCs and RAs.

The CEPT maintains the so-called European Common Allocation Online Database, which is hosted on the ERO website [ERO07b]. Basically, this is an excerpt from the RR including all relevant footnotes. It was cast into the form of a Web interface. So, it is very easy and straightforward to extract information about the spectrum usage in Europe within a given frequency band. Furthermore, there is also information coming from investigations carried out by ERO on the expected spectrum usage in Europe beyond 2008.

3.3 European Commission

In 2002, the European Parliament issued a decision on a regulatory framework for radio spectrum policy in the European Community, which goes under the name of "Radio Spectrum Decision" (RSD) [RSD02]. It introduced a new policy and legal framework in order to ensure coordination of policy approaches and, where appropriate, harmonized conditions with regard to the availability and efficient use of radio spectrum necessary for the establishment and functioning of the internal market in community policy areas, such as electronic communications, transport, and R&D. Its objective is to provide the means for technical implementation of any community policies that are linked to spectrum usage issues. To this end, a committee, namely the Radio Spectrum Committee (RSC) has been set up to assist the EC in the process of developing and adopting these technical implementation measures. Basically, this refers to defining the technical parameters and constraints under which a harmonized spectrum usage amongst Member States should be envisaged. The RSC consists of representatives of the Member States and is chaired by a representative of the EC.

In case, harmonization cannot be dealt with on a purely technical level, the EC may submit a proposal to the European Parliament and to the Council. It was explicitly noted in the RSD that any radio spectrum policy cannot be based only on technical parameters but also needs to take into account economic, political, cultural, health, and social considerations. Moreover, the ever increasing demand for the finite supply of available radio spectrum very likely will lead to conflicting pressures to accommodate the various groups of radio spectrum users in sectors such as telecommunications, broadcasting, transport, law enforcement, military, and the scientific community. Therefore, EC and Parliament request radio spectrum policy to take into account all sectors and to balance the respective needs.

A decision of the European Parliament is binding for Members States according to community law. There are only a few exceptions from this and usually will lead to either transitional rules or sharing mechanisms for those telecommunication services that are affected by the European harmonization process. To take account of the various different interests of commercial and noncommercial spectrum users the European Commission may base its decisions on the outcome of public consultations.

The primary objective of the RSD is to manage the spectrum usage across Europe in a harmonized manner. It is not intended to cover frequency assignment or licensing procedures on a national level. The RSC is expected to cooperate with experts from national authorities in the field of spectrum management, in particular collaboration with the CEPT is foreseen. Usually, this is based on a mandate issued by the Commission to the CEPT asking to elaborate on certain aspects of spectrum usage. In the first place, the RSD aims to make the use of radio spectrum more flexible and ensure the development of a single European market for equipment and services.

This philosophy is reflected by the second annual progress report on the RSD in 2005 [Com05a], which proposed a market-based approach to spectrum management in Europe [Com05b]. In particular, the European Commission has proposed that specific bands—spectrum used for "electronic communications services"—should be subject to tradability throughout the EU. Spectrum trading means buying or selling the right to use a frequency band. The idea is that trading can help to determine the "market value" of spectrum, so the introduction of this approach would help reconcile demand and supply. Furthermore, the European Commission considers spectrum trading a tool to drive innovation and the development of new wireless technologies.

In addition to the RSC, there is another group dedicated to spectrum issues. This is called Radio Spectrum Policy Group (RSPG). It was established as one of the actions following the adoption of the RSD. The RSPG shall adopt opinions, which are meant to assist and advise the European Commission on radio spectrum policy issues, on coordination of policy approaches, and on harmonized conditions with regard to the availability and efficient use of radio spectrum. The members of the group are representatives of the Member States and of the Commission. Furthermore, several observers, for example from the European Parliament, CEPT, and ETSI, attend the meetings of RSPG as well. The European Commission expects the RSPG to consult extensively and in a forward-looking manner on technological, market, and regulatory developments relating to the use of radio spectrum in the context of EU policies on electronic communications, transport, and research and development. Such consultations should involve all relevant radio spectrum users, both commercial and noncommercial, as well as any other interested party.

3.4 National Authorities

The previous sections were dealing with the three major players in the European spectrum management business. However, it is clear that underneath the international level there are national authorities in each country, who manage the spectrum usage on a national level. Their activities are subject to national laws. On the other hand, all national policies and decisions are embedded into the framework set up by ITU, CEPT, and EC. In case neighboring countries are affected by spectrum usage in a particular country, bi- or multilateral negotiations and coordination is required.

European countries exhibit significant differences in relation to policies and economics. Therefore, national laws concerning the spectrum usage are very different, too. There are countries, which have very centralized administrative structures such as France and United Kingdom, for example. But, nevertheless, they follow different policies, for example, United Kingdom puts more emphasis on market-oriented utilization of the electromagnetic spectrum while in France this is more strictly regulated. In Germany, as a complete different example, the situation is more complex. Broadcasting is subject to federal law. Therefore, the licensing of frequencies for broadcasting in Germany has to undergo a two-fold process. Whether a given frequency or channel can be used for broadcasting is decided on a national level, while the details of the content are subject to federal policies.

3.5 Frequency Allocations in Europe

Following the World Administrative Radio Conference in 1992 (WARC-92) CEPT endorsed to pursue the adoption of a harmonized "European Table of Frequency Allocations and Utilisations" by the year 2008. This table has been assembled by ERO and was based on three different "Detailed Spectrum Investigations" (DSI)[5] dealing with distinct frequency ranges and carried out between 1992 and 2000. One of the objectives of the corresponding surveys was not only to update the actual spectrum usage in Europe, but also to identify potential future harmonized spectrum usages across Europe.

---

[5]More information on DSI can be found on the ERO website [ERO07a].

The resulting "European Common Allocation Online Database" (ECA) on the ERO Website [ERO07b] allows to very easily retrieve information about the spectrum usage in Europe. It contains information taken from the latest version of the RR [ITU04] as well as the DSI outcomes on current and future spectrum usage in Europe. Therefore, it is possible to gain an overview about the frequency bands allocated to broadcasting services and, moreover, identify those frequency bands where sharing with other telecommunications services is necessary. Table 3.1 summarizes the frequency allocations in bands I - IV and L-Band, i.e. the frequencies 45–85 MHz, 87.5–108 MHz, 174–230 MHz, 470–606 MHz, 606–862 MHz and 1452–1479.5 MHz as well.

Table 3.1: Frequency allocation within RR and CEPT.

Frequency [MHz]	RR	ECA
Band I		
46.4–47	Fixed, Mobile	Mobile
47–50	BC	Land mobile
50–52	BC	Land mobile, amateur
52–68	BC	Land mobile
68–70.45	Fixed, Mobile*	Mobile
70.45–74.8	Fixed, Mobile*	Mobile*, radio astronomy
74.8–75.2	Aeronautical radionavigation	Aeronautical radionavigation
75.2–85	Fixed, mobile*	Mobile
Band II		
87–108	BC	BC

Band III		
174–216	BC	BC, Land mobile
216–223	BC	BC
223–225	BC, fixed, mobile	BC
225–230	BC, fixed, mobile	BC, land mobile
230–240	Fixed, mobile	Mobile
Band IV		
470–608	BC	BC, mobile
Band V[6]		
608–614	BC	BC, mobile, radio astronomy
614–790	BC	BC, mobile
790–838	BC, fixed	BC, mobile
838–862	BC, fixed	BC, mobile
L-Band		
1452–1479.5	BC, BC-satellite, fixed, mobile*	BC, BC-satellite, fixed, mobile*

Note: "BC" means broadcasting and the asterisk indicates mobile allocations except aeronautical mobile:-

---

[6]The entries of the Frequency Table of Allocations of RR refer to the decisions of WRC-03 [ITU03]. At WRC-07 that was held in autumn 2007 in Geneva a primary allocatoin of mobile services in the frequency range of 790–862 MHz has beed agreed for ITU-Region 1. At the time when this book was written the updated RR were still to be published.

Chapter 4

Coverage Prediction

Digital terrestrial broadcasting systems employ very highly developed technologies. Their technical characteristics can be adapted to very different conditions. Network operators as well as content providers can choose from a great variety of distinct operation modes the one that complies the best with the envisaged coverage target. Highly sophisticated error protection mechanism are integrated to reduce the impact of bad propagation conditons might have to a minimum. The overall data rate required to provide a satisfying quality of service to the customer has been increasingly reduced in recent years by the introduction of more efficient source coding algorithms.

Confronted with such a complexity it is evident that the process of bringing into operation a digitial terrestrial broadcasting network is not an easy task. Basically, this process consists of two steps, First, each terrestial network is in need of an appropriate frequency or channel that could be used. This touches on the problem of frequency assignment to different terrestrial broadcasting services. Typically, a set of different services is to be provided throughout a given geographical area. Clearly, different frequencies are then required for each of the terrestrial broadcasting service in order to avoid mutual interference between them. Second, a transmitter network has to be planned that actually uses the available frequency in order to distribute the audio or video programs. Depending on the technology of the terrestrial broadcasting system a network can consist of a single transmitter or a set of transmitters that conjointly deliver one or more programs within a given area.

As discussed in Chapter 2, the characteristics of a particular digital terrestrial broadcasting system are determined by a large number of different technical parameters. However, it turns out that for frequency planning and network planning most of the technical details are not relevant. Fortunately, both frequency and network planning can be modeled in terms of a limited set of parameters. Nevertheless, it has to be noted that this reduction in number of relevant parameters does not imply that frequency and network planning are straightforward tasks. Quite the contrary, the mathematical problems associated with are usually so involved that highly sophisticated mathematical algorithms need to be exploited to obtain satisfying results.

Both frequency and network planning rest on the possibility of carrying out proper service or coverage predictions. Basically, this means that in the first place methods are required that allow to predict the field strength a single transmitter produces at a particular geographical location. To this end, the technical characteristics of the transmitter, the receiving conditions and also the wave propagation mechanisms need to be known and appropriately taken into account. There are several prediction models that are used in general. Their basic ideas will be explained in the following.

A transmitter located at a given geographical site emits an electromagnetic wave giving rise to a certain field strength level at a given point of reception. The magnitude of the field strength level depends on the technical characteristics of the transmitter such as the radiated power and the height at which its antenna is mounted. Clearly, other parameters such as the distance between transmitter site and point of reception or the topographic conditions between transmitter and receiver are important as well. In particular the latter, namely topography, combined with morphologic conditions predominantly determines the level of the field strength to be expected at a given point. All these impacts are, however, dependent on the wave length of the radio signal.

Predicting a certain field strength at the receiving site is just the first step on the way to assess the quality of service at the point of reception. System aspects need to be taken into consideration as well. This refers to a minimum field strength that is to be provided in order to allow the receiver to decode the information contained in the received signal. In the presence of interfereing signals, the ratio between the wanted signal and the interfering signals need to exceed a certain threshold as well. Otherwise the receiver will not be able to distingiush the wanted signal and the noise and/or interference background. Clearly, these features differ from system to system.

Wave propagation prediction combined with relevant systems parameters allows to anticipate the quality of service at a considered point of reception. However, it is obvious that it is not sufficient for the assessment of a broadcasting network to give evidence about a single point. Network providers need to know the quality their networks actually offer basically at every point within a given area. From a mathematical point of view this not feasible due to the fact that the prediction calculations cannot be carried out by analytical mathematical methods. Rather, numerical computations are needed with the help of computers. Consequently, the claim "at every point within an area" needs to be understood in terms of "a finite set of distinct points appropriately close to each other within an area".

The fact that only for distinct points an assessment of the service quality can be practically carried out is compensated for by employing statistical methods. Usually, based on physical facts assumptions are made about how the field strength values statistically vary in the vicinity of a given point. Thereby, it is possible to calculate the quality of service within a small area around the point of reception in a probabilistic sense.

4.1 Technical Characteristics of Transmitters

For the purpose of frequency assignment or network planning a real transmitter needs to be described in mathematical terms. Clearly, the geographical location where the terrestrial transmitter is situated is crucial for its impact on the service quality. Depending on the geographical altitude of a transmitter site different areas can be served. In mountaineous regions the propagation of electromagnetic signals is hindered by obstacles between the transmitter site and a point of reception.

The radiated power of a transmitter is the most important technical parameter. But, this is directly linked to the antenna used for the emission. Every antenna gives rise to a characteristic three-dimensional angular antenna pattern. The most simple antenna pattern is that of a three-dimensional isotropic radiator. In that case, the radiated power is released uniformly in each direction. However, spherical symmetry is not desirable for broadcasting purposes. The transmission should preferably take place in horizontal direction. A slight tilt towards the earth's surface is even better.

With the help of proper antenna design it is possible to bundle the available power towards defined directions. This nonisotropic transmission is described in terms of the so-called antenna gain. It is defined as the ratio between the power density of the boresight direction of the considered antenna and the power density of a ideal half wave dipole.

The antenna gain is a global characterization of an antenna. Basically, it is a first indication for the directivity of the antenna. To provide the full information about the angular behavior it is, however, necessary to make reference to the complete antenna pattern or diagram.

In addition, the height above ground at which the antenna is mounted at the transmitter mast crucially determines the field strength at a given point of reception. If an antenna is mounted only a few meters above ground, it is evident that even large buildings might attenuate the emitted signals significantly. Therefore, the higher the antenna is located at the mast the larger the range of the transmison will be in general. The technical characteristics of a terrestrial transmitter are discussed in more detail in Section 6.1.1.

4.2 The Terrestrial Radio Channel

The relevant frequency ranges for the terrestrial transmission of radio and television lie between 30 MHz and 1.5 GHz in Europe. Across such a large frequency range the wave propagation conditions are varying heavily. On their path from the transmitter to the receiver the electromagnetic signals are affected by the topographical and morphographical nature of the earth's surface. At a first approximation, the wave propagation takes place quasi-optical. Surely, this assumption complies better for high frequencies than for the frequencies at about 100 MHz.

At the point of reception, very often the resulting signal consists of a superposition of several different signal contributions. If there is a line-of-sight connection between the transmitter and the receiver location then the direct signal makes up the largest contribution. Terrestrial radio and television transmitters are usually set up at locations high above sea level. Their transmitting antennas are preferably mounted high above ground on a high mast, whereas the reception antenna is in general only a few meters above ground. This geometry results in so-called ground reflections from the immediate vicinity of the receiver. They make a large contribution to the resulting signal. Typically, they are of the same order of magnitude as the direct signal. Furthermore, signals emanating from the transmitter location will be scattered or reflected by mountains, hills, or buildings. Thereby additional echoes are generated arriving at the receiver site with characteristic time delays. The full set of these different signal contributions constitutes the already mentioned multipath environment (see the discussion in Section 2.1).

The details of the transmission characteristic mainly depend on the radio frequency. The higher the frequency the smaller structures can in principle act as scattering centers for the electromagnetic waves. In the range of 1.5 GHz, which is one of the frequency ranges where for example T-DAB is operated, the wave length is about only 20 cm. Each mast of a traffic light, each traffic sign, or the metallic window frames in multistory buildings should in principle be included in the description of the transmission. It is quite obvious that this is not possible since this would require an enormous amount of data to be handled provided this detailed information is available at all.

To aggravate the situation, many potential scattering centers are not fixed. This includes in the first place, cars. But also in the case of indoor reception there are moving scattering centers, namely people moving around. Therefore, the transmission channel is time variant. It is obvious that this kind of variance based on erratic movement of cars and people can only be taken into account in a statistical manner. Consequently, the temporal fluctuations of the composition of a set of contributing waves at a particular point of reception and thus the resulting field strength can also be described in statistical terms only.

However, measurements of typical transmission scenarios fortunately show that in practice, in most cases, the situation is quite stable. At a chosen point of reception, usually only a couple of signal contributions significantly add to the resulting field. In the first line, this comprises the direct signal and the ground reflections. For analogue systems that typically use a smaller bandwidth than digital systems, this sometimes creates severe problems because these two components are of comparable strength. Their relative time delay is very small, and thus the degradations in the spectrum might be fatal (see Chapter 2.1). As a consequence, the receiver no longer can demodulate the signals and retrieve the transmitted information without errors.

In addition to direct path and ground reflection, there are also reflections caused by geographic obstacles. Nevertheless, the number of significantly contributing echoes is limited in practice. Figure 4.1 shows

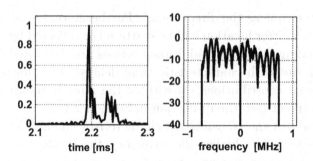

Figure 4.1: Measured impulse response (on left side) and associated transfer function (on right side) taken from the
VHF T-DAB in Baden-Württemberg, Germany.

the result of one particular measurement taken in the country of Baden-
Württemberg in Germany in the running T-DAB transmitter network.
It refers to a T-DAB network that is operated in VHF using T-DAB
block 12B.

At the location where the measurements were carried out, two transmitters of the single frequency network could be received. Since the
two transmitters were separated from the point of reception by different
distances the corresponding signals had different times of arrivals. Indeed, there are two groups of signals to be seen on the left picture of
Figure 4.1 where the impulse response function of the radio channel is
depicted. Both groups consist of a direct signal and a number of echoes
that are caused by reflections somewhere on the way from the transmitter
to the receiver. The right picture of Figure 4.1 shows the corresponding transfer function that has been calculated from the impulse response
function by applying the FFT.

What attracts the attention are the periodic fadings in the spectrum.
They are a direct consequence of the presence of the two signal groups. If
the measured data are used to determine the distance between two min-
ima in the spectrum a value of $\Delta f \simeq 118$ kHz is found, which according
to the rules of the Fourier transformation leads to a temporal separation
of two localized structures in the impulse response function of about 0.05
ms. This is approximately equal to the time difference between the two
groups in the impulse response picture. Following the terminology of
Section 2.1 Figure 4.1 illustrates a frequency selective fading situation

because

$$\Delta f = B_C \simeq 118 \ kHz << B = 1.5 \ MHz \ . \tag{4.1}$$

The description of a transmission channel by means of the phrases "impulse response" and "transfer function" implies a linear dependence between received signal $s(t)$ and the transmitted signal $\overline{s}(t)$. Nonlinearities that might be important for different propagation mechanism do not have any relevance here.[7] However, it has to be noted that linear does not necessarily mean proportional. Rather, the general link between transmitted and received signal is given as

$$s(t) = \int dt' \ R(t - t')\overline{s}(t') + n(t), \tag{4.2}$$

where $R(t)$ represents the impulse response function and $n(t)$ is an additive noise term. Equation (4.2) says that the received signal s at the time t depends on all values of the transmitted signal $\overline{s}$ for all times t'. Clearly, in reality the causality between transmitting and receiving must be kept. Only signals that have been broadcast before time t can have a physical influence on the received signal.

Impulse response $R(t)$ and transfer function $\Gamma(\omega)$ can be transformed into each other with the help of the Fourier transformation. In the spectral regime Equation (4.2) thus is equivalent to

$$\xi(\omega) = \Gamma(\omega)\overline{\xi}(\omega) + \eta(\omega), \tag{4.3}$$

where $\eta(\omega)$ is the Fourier transform of the noise term $n(t)$. The quantities $\overline{\xi}(\omega)$ and $\xi(\omega)$ are the Fourier transforms of the transmitted and the received signal, respectively.

The impulse response function that can be seen in Figure 4.1 is valid in that form only at one single point of reception. In the vicinity of this point the transmission characteristics might change dramatically. A very simple example will shed some light on this. If two equally powerful transmitters can be received under a relative angle of incidence of 30° between them a very typical interference pattern is generated by the superposition of the two signals within an area with a dimension of several wave lengths. Such a wave field is shown in Figure 4.2.

---

[7]In fiber cable nonlinear propagation modes become more and more important. They can be excited by using very high power ultra-short lasers pulses. Then the nonlinear impact of the refraction index on the propagation comes into play giving rise to nonlinear propagation modes called solitons.

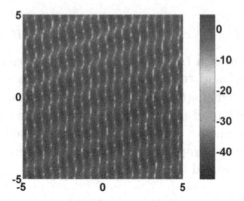

Figure 4.2: Wave field in the vicinity of a chosen point of reception that is assumed to be located in the center. The amplitudes are given in dB and normalized with respect to the maximum value of the field. The distances are multiples of the wave length.

Now, if the receiver starts to move from one point to another, variations of the field strength level up to 20 dB are experienced. Sometimes at critical locations the changes can be in the order of 40 dB. These fluctuations of the field strength in an area of the order of several wave lengths are usually called fast fading. Its origin is of purely physical nature, namely the interference of a set of signals.

In reality the structure depicted in Figure 4.2 is time-variant, too. Even when the receiver is fixed at one particular location the wave field is changing in the course of time, depending on how the composition of the set of arriving signals changes. There might be changes in the number of incoming waves, their amplitudes, their angles of incidence and their relative time delays. These changes cannot be described by deterministic means. Only in very few and very special situations, this is possible.

In most cases, the description of the terrestrial radio transmission channel is based on statistical approaches in order to model the spatial and temporal behavior of the transmission characteristics adequately. Against this background numerous channel models have been developed from which some emerged being more relevant for the planning of terrestrial transmission systems than others. The four most important models will be presented here very briefly, just to summarize their basic features. For a more profound discussion, it is referred to standard text books of transmission technology such as [Pro89] or [Kam92].

4.2.1 Gaussian Channel

In the literature, the Gaussian channel is very often called 'AWGN channel'. The capital letters stand for '*A*dditive *W*hite *G*aussian *N*oise'. This refers to a transmission scenario, where the signal at the point of reception is given as the superposition of the transmitted signal and a noise term according to

$$s(t) = \overline{s}(t) + n(t) \ . \tag{4.4}$$

The impulse response function consequently reads

$$R(t) = 1 \tag{4.5}$$

according to the general formulation (4.2). The quantity $n(t)$ is assumed to be Gaussian noise, that is $n(t)$ has to be considered as a statistical variable which follows a Gaussian probability distribution with zero mean and standard deviation σ_n. Hence $s(t)$ has to be treated as a statistical variable, too, with Gaussian distribution according to $n(t)$.

For terrestrial broadcasting the Gaussian channel is only of subordinate value in practice. The fact that theoretical investigations again and again refer to the Gaussian channel is connected to its mathematical simplicity. Only this very simple model allows the closed, analytical calculation of quantities like bit error ratios at tolerable costs. The advantage of closed form expressions is obvious. Systematic investigations of the impact of certain parameters on the transmission quality can be carried out easily with their help. In all the other cases, it is necessary to fall back on very time-consuming numerical simulations to come to reliable conclusions.

Another advantage of the simplicity of the Gaussian channel has to be mentioned. More complex channel types normally require higher field strengths at the point of reception, to guarantee a certain reception quality. If direct calculations on the basis of corresponding channel models cannot be afforded it might be possible, however, to give gross estimates, for example, of the protection requirements of a complex transmission scenario by adding margin values to the results derived for the Gaussian channel.

4.2.2   Stationary Multipath Channel

Every time not only one but several signals also superimpose at a point of reception and this constitutes a multipath channel. In contrast to the analogue world, the wanted signal contributions in digital single frequency networks can come from different transmitters. The signals broadcasted by each of the transmitters may undergo different propagation effects, leading to a great variety of different signal contributions arriving at the point of reception.

The most simple way to illustrate such a scenario is by employing sinusoidal waves. The m-th transmitter is assumed to broadcast the signal

$$\bar{s}_m(t) = \exp\left[i\omega t + \varphi_{m0}\right] \tag{4.6}$$

where φ_{m0} is an arbitrary initial phase and ω the frequency. According to the channel characteristics each of the signals has to undergo, the received signal is to be described by the linear superposition

$$s(t) = \sum_m \sum_k A_{mk_m} \exp\left[i\omega\left(t - \tau_{mk_m}\right) + \varphi_{m0}\right] + n(t) . \tag{4.7}$$

Any noise effects are accounted for by adding the noise term $n(t)$. The parameter A_{mk_m} represents the attenuation factor the k_m-th path originated from transmitter m has suffered. Its time delay is given by τ_{mk_m}. Equation (4.7) can be cast into a form that is in accordance with relation (4.2). The impulse response function thus reads

$$R(t) = \sum_m \sum_k A_{mk_m} \delta\left(t - \tau_{mk_m}\right) . \tag{4.8}$$

In (4.8), $\delta(t)$ denotes the delta function. The associated transfer function is given by

$$\Gamma(\omega) = \sum_m \sum_k \tilde{A}_{mk_m} \exp\left[-i\omega\tau_{mk_m}\right] + \eta(\omega) . \tag{4.9}$$

The constants $\tilde{A}_{mk_m}$ and A_{mk_m} differ only by some numerical factors that depend on the employed definition of the Fourier transformation.

The radio channel defined in Equation (4.8) is assumed to be not time-variant, that is all parameters involved do not show any time dependence. In practice, a Gaussian noise term $n(t)$ is used. Such types of multi-path channels are typically applied for numerical simulations or also for measurements in the laboratory. In [Com88] four channel models for mobile telecommunication systems such as GSM have been defined. These models were applied to investigations for T-DAB and DVB-T very often.

4.2.3 Rayleigh Channel

In both the preceding sections, it was presumed that the parameters of the radio channel are not time-variant. However, in practice this is an exceptional case. Normally, the channel parameters vary in time. Both scattering as well as reflection of the electromagnetic waves are affected by fluctuations. Furthermore, mobile reception as a typical reception situation naturally leads to time dependencies. Consequently, the definition of the impulse response as given by (4.8) must be modified such as to take into account time-dependent channel parameters. This gives the representation

$$R(t) = \sum_m \sum_k A_{mk_m}(t) \; \delta\left(t - \tau_{mk_m}(t)\right) . \qquad (4.10)$$

The time dependence of the the amplitudes A_{mk_m} and the time delays τ_{mk_m}, unfortunately, is not known. Furthermore, it is rather likely that also the number of signals is changing in time. The impulse response function thus might change its structure drastically.

Without any knowledge about the detailed dynamics of the channel parameters there remains no other possibility but to interpret these quantities as statistical variables. If the number of independent signal contributions that superimpose at the point of reception is sufficiently large, then it is permissible to treat the resulting received signal as a Gaussian statistical variable. This approximation gets better as the number of contributing signals becomes larger. In the mathematical literature this is known as the central limit theorem.

The superposition of a large number of complex valued plane waves according to an expression such as (4.7) therefore, means that $s(t)$ is complex valued, too, that is

$$s(t) = s'(t) + i * s''(t) . \qquad (4.11)$$

The Gaussian assumption, hence, implies that both the real part $s'(t)$ and the imaginary part $s''(t)$ have to be considered as uncorrelated statistical variables that independently follow a corresponding Gaussian probability distribution function. For the magnitude of $s(t)$,

$$|s(t)| = \sqrt{s'^2(t) + s''^2(t)} , \tag{4.12}$$

a so-called Rayleigh distribution can be derived. It can be cast into the form

$$p(x) = \begin{cases} \dfrac{2x}{\sigma^2} \exp\left[-\dfrac{x^2}{\sigma^2}\right] , & x \geq 0 \\[2em] 0 , & \text{else} . \end{cases} \tag{4.13}$$

The parameter σ stands for the standard deviation while σ^2 gives a value for the received power. The mean value of the Rayleigh distribution is given by $\frac{\sigma}{2}\sqrt{\pi}$. The phase $\varphi(t)$ of the received signal is equally distributed across the interval $[-\pi, \pi]$.

Situations that can be satisfactorily characterized by a Rayleigh channel are encountered if there is no direct line-of-sight connection between transmitter and receiver. Furthermore, it is necessary that the direct path suffers so much attenuation by diffraction that it does not contribute significantly to the resulting received signal. The second group of signals that can be seen in the impulse response function in Figure 4.1 could act as an example. However, it has to be noted that in this case, probably, the assumption for the application of the Gaussian channel is violated, because there is only a very limited number of individual signals involved.

4.2.4   Rician Channel

Very often the direct signal does not suffer such an attenuation that it can be neglected. In most cases, there is one strong contribution the amplitude of which does not fluctuate much. The time of arrival also has a constant value. In single frequency networks, it might happen that several such contributions arising from different transmitters can be received. In addition to these more or less static signals, there is large number of signals having pronounced smaller amplitudes. Under the assumption that both large peaks in the impulse response of Figure 4.1 do not change significantly in time, this would constitute a nice example for a Rician channel.

The power ratio between the direct and the scattered components is called Rician factor c. As in the case of the Rayleigh channel the magnitude of the received signal $|s(t)|$ is a statistical variable following the probability distribution function

$$
p(x) = \begin{cases} \dfrac{2x}{\sigma^2} \exp\left[-\left(\dfrac{x^2}{\sigma^2}+c\right)\right] \ I_0\left(\dfrac{2x}{\sigma}\sqrt{c}\right) \ , & x \geq 0 \\[6mm] 0 \ , & \text{else} \ . \end{cases} \tag{4.14}
$$

The quantity I_0 denotes the modified Bessel function of order zero. The function given by (4.14) is called Rician distribution.

The phase of the received signal is not equally distributed any more as in the case of the Rayleigh distribution. A shape of the distribution for the phase $\varphi(t)$ can be derived, which at first glance is quite similar to an ordinary Gaussian distribution function. The larger the Rician factor c becomes, the more localized the phase distribution gets. This behavior comes as no surprise. An increasing Rician factor means that the echoes are getting smaller and smaller with respect to the amplitudes of the directly incoming signals. If the direct signals, indeed, have constant amplitude and phase, then the impact of the additional contributions is vanishing and both the distribution for the magnitude and the phase of the received signal must converge to a pure delta function in [Kam92].

4.3 Wave Propagation Models

The detailed knowledge about the physical characteristics of the terrestrial radio channel is essential information required to assess the coverage situation within a particular area. However, this knowledge must be cast into a form that allows to predict the field strength level at a given point of reception. To this end, an appropriate prediction model is needed. A large number of such wave propagation models has been developed over the years, most of them adapted to different propagation conditions.

Seen from a physical perspective, free space propagation, reflection, diffraction, and scattering are the physical mechanisms of electromagnetic wave propagation that determine the terrestrial transmission of radio and television signals. The number of publications extensively

dealing with this topic is nearly inexhaustible. As introductive literature the two text books [Hal96] and [Hes98] could be recommended.

The calculation of the field strength at the receiver location presumes
sufficient knowledge about all relevant topographic and morphologic conditions that might affect the propagation of the electromagnetic waves
on their way from the transmitter to the receiver. Basically, topography refers to the height above sea level of a given point while the term
morphology covers the surface structure at or in the vicinity of the point
of reception, that is whether that point lies within a densely populated
area, a rural area or if it is on farmland, forest or sea. These boundary conditions have a significant impact on the wave propagation. The
topographical and morphological information is usually stored in corresponding data bases.

Practically, such data can be sampled only with a limited spatial
resolution. It is not possible to define a height function such that for
any arbitrarily chosen geographical point a height over sea level can
be derived or calculated. Usually, such information is only available
for points that are separated by a distance d. Therefore, the earth's
surface is approximated by plane squares with an edge length equal to
d. Each square is assigned the height above sea level that corresponds
to the grid point lying there. However, this means that the height value
originally associated with a single point now represents a square area.
In a similar way, a square area approximation of the earth's morphology
can be drawn up.

The choice of the value of the resolution d depends both on the accuracy of the available data and on the wave length of the considered
radio signals. The smaller the wave length becomes, that is the higher
the frequency is, the smaller the dimensions of obstacles that can potentially affect the wave propagation become. The distance should be
small enough to properly take into account the effect of topography and
morphology on the propagation.

All wave propagation prediction methods have to struggle with the
fact that the radio channel is time-variant in practice. The field strength
level is fluctuating more or less heavily with respect to a mean value. As
discussed in the previous sections, this fluctuations can only be described
statistically. Hence, strictly speaking no deterministic field strength F
can be calculated. Rather, the result of any calculation has to be interpreted in terms of saying the field strength at the point of reception
under consideration exceeds the value F in $x\%$ of the time. Typical val-

ues of x are 1%, 10% and 50% (see e.g. [Hal96] or [Hes98]). Figure 4.3 demonstrates the interpretation of a calculated field strength value.

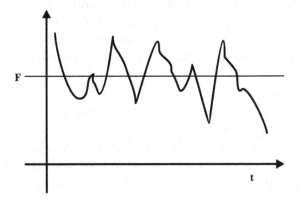

Figure 4.3: Interpretation of calculated field strength value F as the value that will be exceeded in x% of the time. In this case $x \approx 50$%.

To avoid any misunderstandings, it has to be emphasized that the term field strength has a very distinct meaning in the context of wave propagation models. Physically, the impact of an electromagnetic wave at the location of the receiver is to deliver a certain power P, which is linked to the physical electromagnetic field strength of the wave E according to $P \sim |E|^2$. However, it is common ground in the context of wave propagation models employed for frequency and network planning to use the terms "field strength" or "field strength level" for the logarithm of the normalized power P of the wave, that is

$$F = 10 \log \frac{P}{P_0} = 10 \log \frac{E^2}{E_0^2} \qquad (4.15)$$

where E_0 corresponds to $1\,\mu\mathrm{V/m}$. All wave propagation models refer to the quantity F, that is they give field strength values in units of $\mathrm{dB}(\mu V/m)$.

4.3.1 Recommendation ITU-R P. 1546

If there is only little information about the topography of a region for which a field strength prediction is to be carried out, then the model of the ITU based on the recommendation ITU-R P.1546 [ITU01a] offers a way to make a field strength prediction. The recommendation ITU-R P.1546 replaced the formerly valid recommendation ITU-R P.370, which has been used for decades till the end of 2001. Recommendation ITU-R P.1546 (or its predecessor at the respective time) is the standard wave propagation model for broadcasting because international frequency planning conferences and any coordination activity under ITU custody are based on this model.

Topography is not taken into account explicitly. In contrast, the model is based on a vast number of field strength measurements taken over a period of many years. This information has been condensed into a number of propagation curves from which the field strength value at a chosen distance from a transmitter can be extracted. Clearly, the propagation curves are valid only for a predefined set of special transmitter characteristics, for example, they assume a transmitter output power of 1 kW.

It has to be kept in mind that the calculated field strength values have to be interpreted in a very particular way. First of all, the values are to be understood as a value exceeded by the physical field by a certain amount of time as described in the previous section. Second, due to the derivation of the curves from measurements it is very unlikely that a calculated value for a certain point of reception will be reproduced when making a measurement of the field strength there. The calculated value is valid only in a statistical way in the sense that if a lot of measurements would be carried out at the same distance from the considered transmitter but in arbitrary directions, then the mean of these measured values should coincide with the calculated figure.

At first glance the method based on the ITU recommendation ITU-R P.1546 seems to be applicable only on very few special occasions due to the fact that propagation curves are given only for a limited set of technical parameters. This is clearly not true since at the same time interpolation rules are given to apply the method for other parameters as well. In principle, any transmitter and receiver site characteristics can be taken into consideration. This covers arbitrary output powers of the transmitter as well as, for example, different heights of the receiving antenna above ground.

In order to grasp the essential technical features both at the location of the transmitter and the receiver site, which can influence the wave propagation, a set of special technical parameters has to be provided for the application of the method. The most important parameters are the time percentage associated with the field strength level, the distance where the field has to be calculated, the height of the transmitting antenna, and the frequency of the signal.

The field strength curves of recommendation ITU-R P.1546 are given for the three time percentage values 1%, 10% and 50%. With respect to the distance, 78 different values between 1 km and 1000 km are included. For the antenna height eight nominal values are taken into consideration reaching from 10 m to 1200 m. The frequency range covered by the recommendation reaches from 30 MHz up to 3000 MHz. However, only for 100 MHz, 600 MHz, and 2000 MHz field strength curves are quoted.

Apart from these four essential parameters there is also a categorization of the general propagation type. Since the wave propagation conditions above land, cold sea, and warm sea differ dramatically these three propagation regimes are distinguished. Cold sea refers to the Atlantic Ocean, while warm sea is connected to the Mediterranean Sea. Warm sea is to represent, for example, the Red Sea.

All curves are provided both graphically and tabularly. The latter form is important if the characteristics of a transmitter under consideration do not match with those values given in the recommendation. For that instance interpolation rules and formulas are given. In most cases, it is necessary to make several interpolations. To this end, not only the interpolations, but also rules about the sequence of their application are presented. Both the tables as well as the interpolation rules are innovations, which have been introduced when migrating from recommendation ITU-R P.370 to ITU-R P.1546. The application of the 370 version very often led to problems during application in the past due to the lack of precise rules for interpolation.

The impact of topography is mainly covered by the height h_1 of the transmitter. Its definition is slightly broader in recommendation ITU-R P.1546 than in the former recommendation 370. Depending on the distance in which the field strength is to be calculated, different rules apply. In order to calculate the parameter h_1 first of all the height of the antenna center above ground h_a has to be known. Furthermore, the effective height h_{eff} of the transmitter with respect to the receiver location has to be known. This concept has been taken over from the former recommendation ITU-R P.370. It is defined as being the difference between h_a and the average value of the ground level height calculated from an interval of 3 km and 15 km in the direction towards the receiver. Figure 4.4 visualizes the definition.

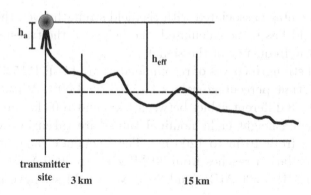

Figure 4.4: Definition of the effective height h_{eff} according to [ITU01a].

The parameter h_1 is defined differently, depending on the distance between transmitter and receiver, namely

$$
h_1 = \begin{cases}
h_a & , \quad d < 3 \text{ km} \\[2em]
h_a + (h_{eff} - h_a) \dfrac{(d-3)}{12} & , \quad 3 \text{ km} \leq d < 15 \text{ km} \\[2em]
h_{eff} & , \quad 15 \text{ km} \leq d
\end{cases} \qquad . \qquad (4.16)
$$

The topographic and morphologic conditions in the vicinity of the receiver are captured by providing different heights R for the receiving antenna height h_2. In order to characterize an inner-city situation the value $R = 30$ m should be used, less densly populated cities are to be represented by $R = 20$ m and rural reception conditions are taken into account with the choice $R = 10$ m. If the actual receiving antenna height does not coincide with one of the reference values, that is if, for example, in a rural area an antenna height $h_2 = 20$ m must be used, then the field strength values have to be corrected following the given interpolation rules.

Since the real topography between transmitter and receiver is not taken into account explicitly, the calculated field strength values sometimes diverge strongly from the actually measured values. This might be

considered as a severe drawback of this wave propagation model. However, the invaluable asset of the ITU-1546 curves is that they represent
some kind of widely accepted international minimal consensus. Nearly,
all frequency planning conferences in recent years would not have come to
any result at all, if the problems connected to wave propagation not had
been dealt with by an ITU Recommendation or a modification thereof.
The application of ITU-R Recommendation P. 1546 [ITU01a] creates
a common basis for all partners participating in the international frequency planning business. When it comes to bilateral negotiations between countries, regions, or even network providers, however, it is true
that more refined models are employed to solve problems on a more
detailed level.

4.3.2 Terrain Based Propagation Models

There exist numerous mathematical methods that provide the means for
a field strength prediction within a particular area for both the VHF and
the UHF frequency range. The work of Longley and Rice [Lon86], Okumura et al. [Oku68], Causebrook et al. [Cau82] and Großkopf [Gro86]
as well as the references found therein offer a profound outline of the
different approaches. In essence, all methods are based on the calculation of the diffraction attenuation caused by obstacles on the way from
the transmitter to the receiver. Which geographical objects are included
is determined by examining a corresponding ground level profile. An
important feature of all these approaches is that they only exploit obstacles that lie on a plane of the profile between transmitter and receiver.
Hence, they are two-dimensional methods by definition.

The field strength at the point of reception is determined by reducing
the free space propagation value by factors derived from the properties
of the relevant topographic and morphological structures. The models
differ in the way these corrections are actually carried out and in which
way morphological data enter the process.

Seen from a physical perspective, diffraction means that the electromagnetic waves manage to get in the geometrical shadow zone of an
obstacle. The basis for all calculations concerning diffraction effects is
the artificial problem of diffraction at an infinitely extended half plane
perpendicular to the direction of propagation. The analytical mathematical treatment of that configuration leads to solutions that can be
expressed in terms of Fresnel integrals (see e.g. [Lue84]). In geometrical

terms this means that in the plane of ground height level an geometrical region, the so-called Fresnel zone, is defined according to a precise construction rule. Any obstacle falling substantially into the Fresnel zone contributes to the attenuation.

In practice, transmitter and receiver are quite often separated by not only one but several obstacles. There exist various approaches to tackle that problem. In most cases, the obstacles are dealt with in terms of subsequently ordered half planes. An exact solution to that mathematical problem is feasible for a moderate number of diffraction edges, however, due to the high numerical effort involved they have virtually no meaning for practical applications.

A quite satisfying approximation is to treat the obstacles sequentially not according to their geographical proximity to the transmitter location but according to their impact on the diffraction attenuation [Mee83]. A general constraint for all methods trying to cope with multiedge diffraction is that none of the attenuating humps between transmitter and receiver must be located near to the transmitter. Similarly, the distance between two obstacles should be more than 2 km. The first case does not play an important role in broadcasting where typically the transmitter sites are chosen on high mountains or hills far enough away from any nearby heights.

Morphological corrections also show up in very different ways across the various methods to determine the diffraction loss. Usually, morphological data have an accuracy and resolution not as good as the topographic data. Therefore, very often additional global attenuations are added to compensate that loss of precision. On the other hand, there are models that combine the morphological data with the digital terrain data in the sense that depending on the morphology the ground height level of a considered point is artificially raised. Typical values are 25 m for forest and 10 m for built-up areas.

Since there is no strict physical method for the calculation of the morphological corrections enormous differences between the results of different approaches are encountered. This makes it very difficult to compare the methods.

4.3.3   3D-Models

The propagation of the electromagnetic waves from the transmitter to the receiver is a three-dimensional physical phenomenon. In general, a multitude of signal contributions arrives at the point of reception from different directions (see Figure 2.1). This multipath environment might lead to a totally different field strength level as in the case with only the direct signal. The set of signals can superimpose constructively or destructively, giving rise to either a higher or lower resulting field strength level.

By reducing the wave propagation model to a two-dimensional ground height profile calculated along the line between transmitter–receiver an error is introduced into the description of the wave propagation that cannot be compensated for by empirical margins anymore. Furthermore, the information how obstacles are oriented with respect to the ground height profile gets lost. It is obvious that there must be a difference if the diffraction edge is indeed perpendicular to the direction of propagation or not.

To overcome these shortcomings several 3D-models have been developed in the past. The publications in [Gro95] and [Leb92] can be used as starting points to enter into that field of research. There are models that are entirely based on ray approximations. Both ray tracing and ray launching algorithms have been applied and adapted to individual needs. Heuristic semiempirical methods are the alternative. They rest on matching a set of empirical factors by comparison with measurements.

In the field of broadcasting where wide area predictions have high priority, so far there does not exist a 3D-model, which achieved real relevance for everyday planning purposes. In the first place, most of the existing models do not produce field strength values that are in line with measured values. Shortcomings in the model assumptions might be the reason for this. Another problem of some of the three-dimensional wave propagation models is connected to their implementation as computer programs. Still their time consumption is not acceptable to allow the application on a daily basis. However, in view of the steadily increasing computer performance this might be a restriction to be overcome in the future.

4.4 Full Area Field Strength Prediction

All wave propagation models discussed in the preceding sections allow to carry out a field strength prediction for individually chosen mathematical points only. Since the radio channel is time-variant this field

strength value has to be interpreted as a value that will be exceeded by
the electromagnetic field in a certain percentage of time. Calculations
can be made for different percentages. For wanted signals usually 50%
time percentage is used, while unwanted signals the percentages 1%, 5%,
or 10% are considered. However, calculations at individual points are not
sufficient in order to assess the preformance of a broadcasting network.
Rather, the quality of service has to be investigated throughout a given
area.

Since as a matter of principle it is not possible to calculate the resulting field strength level for all points within a considered area, it is
quite common to use a regular grid of points. The separation of the grid
points depends on the required accuracy and the availability of topographic and morphological data with sufficient high resolution. Typical
distances between grid points lie in the interval between 100 m and 1000
m for broadcasting applications. As a consequence, each grid point is
surrounded by an area of corresponding size, that is 100 × 100 m or 1000
× 1000 m. These small areas are called pixels. At each of these points
a field strength value is determined with the help of a properly chosen
wave propagation model.

It is well known that the field strength level significantly varies across
the area of a pixel. To a first approximation, the fluctuations of the level
are made up by a slowly changing component that is superimposed by
a rapidly changing component. The slow changes are connected to the
variation of the angle of incidence of the incoming signals when moving
within the pixel area from one location to another. The rapid changes are
due to fluctuations of the field strength level. They occur on a scale in
the order of the wave length. Their physical origin is the superposition of
several individual signals. This phenomenon has been discussed already
in Section 4.2. The slow changing component is usually called "slow
fading" and the rapidly changing effect consequently "fast fading."

In practice, both fluctuations are described in statistical terms. The
statistics of the slowly changing field strength level, that is its variation across the pixel area, is governed by a Gaussian distribution to a
good approximation. According to the relation by Equation (4.15) the
power of the signals follows a log-normal probability distribution (see
e.g. [Hal96]). The signal power connected to the quickly changing component follows to sufficient accuracy a Rayleigh distribution (see e.g.
[Hal96] and the discussion in Section 4.2.3). In practical applications,
these dependencies are usually assumed and both mean and standard
deviation are fixed on the basis of measurements.

Equipped with these assumptions the probability can be calculated that the field strength level exceeds a given value at a predefined percentage of locations within the pixel under consideration. This is called the location probability. To this end, the field strength value calculated by means of the wave propagation model for the considered grid point is employed as the mean value for the distribution of the slow fading for the corresponding pixel. For the Rayleigh distribution of the fast fading component a zero mean value is always presumed. As standard deviations the values found in measurements are used.

For analogue services like FM radio location, probabilities of 50% are employed. However, in the case of digital terrestrial broadcasting systems this value is not considered as being enough to guarantee appropriate service quality. The reason for that is the different transition behavior of analogue and digital systems from good reception to failed reception. Analogue systems exhibit a slow degradation of the service quality when the received field strength is getting smaller, for example, in mobile reception when crossing the border of the service area. In contrast, digital systems show an abrupt — almost digital — transient behavior from good to bad reception. Therefore, the location probability needs to be larger in order to guarantee an appropriate service quality throuhgout an envisaged service area.

4.5 Coverage Assessment

The calculation of the field strength throughout a defined area is only the first step for the assessment of a particular coverage situation. It means that the characteristic features of the transmitter have been appropriately taken into consideration. This refers to the geographical location of the transmitter site, the broadcasting frequency, the antenna diagram, and the output power. In general, both output power and antenna diagram have to be properly designed to comply with constraints imposed by other transmitters using the same or adjacent channels. A further characteristic of the transmitter site directly influencing the range of the wave propagation is the antenna height above ground.

In addition to the transmitter and the wave propagation conditions, the receiving side of the broadcasting chain needs to be taken into account, to properly assess the quality of service. It is necessary to specify how large the reception effort should be a customer is expected to bear at the point of reception. It has to be defined what is the proper receiving antenna height. Is it 10 m above ground, which more or less would correspond to a roof antenna or is it 1.5 m above ground, which typically is considered as being representative for mobile reception? Moreover, the selectivity and the sensitivity of the receiver taken into account during calculations must be specified.

From all these data the minimum required field strength at the location of the receiver can be deduced. Minimum field strength is a synonym for the fact that the receiver needs a certain input power to be able to distinguish the transmitted signal from the noise floor. Using a wave propagation model, then allows to determine those points where the considered transmitter delivers a field strength level above the minimum level.

In practice, there are no isolated transmitters. Usually, there are other transmitters in operation using the same or adjacent radio frequencies. Depending on the distance from the transmitter under consideration, these other transmitters might lead to more or less harmful perturbations. According to the employed type of transmission systems, different protection ratios between the useful signal and the sum of all interfering contributions must be met. For digital terrestrial transmission systems, the situation usually gets even more complicated due to the enormous number of different operation modes that all might demand for special individual protection ratios. The documents [CEP95], [CEP96] and [CEP97] offer a very detailed overview about the situation in connection with T-DAB and DVB-T.

If several different signals have to be considered at a certain point of reception at the same time, the question arises how these signals can be combined into a resulting signal. Seen from a purely physical perspective, the solution to the problem is obvious. At the point of reception, all contribution superimpose according to their amplitudes and phases. So, a full linear superposition of the electromagnetic waves is the result. It is quite evident that such an approach is not feasible on the basis of the available information provided by the wave propagation models at hand. Therefore, other possibilities have to be utilized.

In order to derive a method for the combination of several signals, it is helpful to reconsider what the wave propagation models actually provide and how this information is interpreted. Given a set of transmitters, the result of the wave propagation models is a set of field strength values at the receiver site that are exceeded for a certain percentage of time. They are interpreted as being the mean values inside the area of a pixel, well being aware of the fact that the field strength value varies across the area of the pixel according to the corresponding distribution functions. Thus, a deterministic combination of the quantities does not make any sense.

Indeed, a statistical treatment of the unification of the field strength values is mandatory. The problem of superimposing several electromagnetic waves is hence mapped to the problem to find the probability distribution function for the sum of two or more independent signals. The assumption about independence, however, is not quite correct in reality because different signals are sometimes completely correlated as in the case of a direct signal and a corresponding ground reflection. An adequate and settled treatment of correlation when deriving the distribution function of the sum of signals is still missing and under investigation.

As discussed in the previous section, in practice it is assumed that the distribution functions for the field strengths are either log-normal or of Rayleigh type for single signals. The calculation of the distribution function for the sum of such signals is a time-consuming numerical process. Therefore, solutions based on approximations have been developed.

Especially in single frequency networks it is very often not possible to make a clear distinction between wanted an unwanted signal. Sure enough, all signals arriving within the lap of a guard interval after the first used signal can contribute positively. On the contrary, signals arriving with a delay larger than a full symbol length will only produce negative interference. For those in between the situation is not that obvious a priori. Both wanted and unwanted impact might result (see the discussion in Section 2.1 and Figure 2.3).

When having distinguished wanted from unwanted contributions the resulting probability distribution functions for both sums have to be determined. Since both useful and interfering signals are dealt with in a statistical way this quite naturally carries over to their ratio. Once this is established all information is available to calculate the location probability that allows to label a given pixel as served or not served.

The statement "a pixel is covered or served" can be given only in statistical terms, too. Two conditions have to be met thereto at the same time. First, it has to be checked if the minimum field strength exceeds the required value with a probability of p corresponding to the location probability required for the service under consideration. Second, if the probability that the protection ratio exceeds the required value, is larger than p as well, then the pixel is called served. In other words this

pixel is part of the service area. If the minimum field strength criterion
is met but the protection ratio is not sufficient, the pixel is classified as
interfered. Finally, in case not even the minimum field is strong enough
the pixel is simply not served. The term coverage usually refers to those
pixels that are either served or interfered. The idea is that coverage
refers to that area where in principle service could be provided because
the minimum field strength is exceeded. So if there would be a mitigation
technique those pixels classified as interfered could be changed to served
by surpressing the interference.

It has to be emphasized that this is the traditional definition in broadcasting of served, interfered, and not served. However, other definitions
are conceivable as well. For example, if it is implicitly assumed that
high power transmitter networks are implemented only, the minimum
field strength criterion could be omitted, because it could be expected
that the field strength delivered at each point within a given area is high
enough to meet the minimum field strength criterion anyway. Furthermore, instead of considering two independent probabilities for exceeding
the minimum field strength and the protection ratio a joint probability
function might be considered more adequate.

As already indicated above, the crucial point, however, in the whole
process is the statistical combination of several independent signal contribution. The most important methods at least for practical applications
are discussed in some detail in the sections below.

4.5.1   Monte-Carlo Method

Apart from determining the joint probability density functions of several
independent fields by means of numerical integration, the well-known
Monte-Carlo approach can be exploited. Sure enough, it is the most
accurate of all methods based on some kind of approximation. If the
mean values of all wanted and unwanted contributions as well as their
standard deviations are known it is possible to calculate the shape of the
sought after distribution function for the protection ratio between the
total wanted and unwanted signal by simulation.

To this end, for every signal, a set of random values is generated
on the basis of its corresponding distribution function. The set should
contain a significantly large number of entries. Then, a set of wanted
and a set of unwanted signals are created simply by summing. From this

a histogram of the protection ratio is derived that subsequently can be used to calculate the coverage probability numerically.

Such an investigation has to be carried out for each of the pixels of the considered coverage area. It is obvious that this requires an enormous numerical calculation effort to obtain a coverage assessment for a wide area. In practical applications this is prohibitive.

4.5.2 Power Sum Method

If only a very rough, but nevertheless, fast solution is asked for the so-called "power sum method" can be applied [EBU98]. The underlying idea is very radical in the sense that the statistical nature of the wanted and unwanted signals is neglected completely. Useful and interfering components are only summed in terms of powers and then their ratio is determined.

The link between the mean value of the field strength level $\overline{F}$ and the corresponding power $\overline{P}$ reads

$$\overline{F} = 10 \ dB(\mu V/m) \times \log_{10}(\frac{\overline{P}}{P_0}) \tag{4.17}$$

with

$$P_0 = \frac{E_0^2}{Z_0} \quad , \quad E_0 = 1 \ \mu V/m \ . \tag{4.18}$$

Summing all powers according to

$$\overline{P}_S = \sum_n \overline{P}_n \tag{4.19}$$

leads to the desired mean power sum $\overline{P}_S$, which by applying the inverse relation of (4.18) can be mapped to a corresponding field strength again.

Such a summation is carried out both for the useful and the interfering parts independently. The situation that both wanted and unwanted sums are equal is interpreted as a coverage probability of 50%. In this particular case, the results obtained by application of the power sum method are equal to of typical statistical methods to the first approximation. Other ratios lead to different coverage probability whose deviation from other approaches, however, increases and thus are no longer reliable.

4.5.3   Log-Normal Method

The log-normal method (LNM) provides a mean to calculate an approximation to the probability distribution function of a sum of log-normally distributed not correlated signals. It is based on the assumption that the distribution function of the sum of two log-normally distributed signals is also a log-normal distribution. In practice, two implementations of this method won recognition. Only a brief summary of the methods will be given here. A detailed description of both can be found in the annex of [EBU98].

To clarify the difference between the two types of LNMs it must be recalled that the link between the field strength and the power (4.18) establishes also a link for the associated probability distribution functions of both quantities. It is widely accepted that the statistics of $\overline{F}$ is governed by a log-normal distribution. Relation (4.18) forces $\overline{P}$ to be normally distributed.

In principle, this gives two possibilities to combine several signals statistically. With the help of (4.18) all individual signal field strengths are mapped to powers. The statistics of the sum of powers is straightforward due to the fact that the individual $\overline{P}$s follow a Gaussian distribution, respectively. The only problem is to map back the normal distribution of the power sum to the corresponding log-normal distribution of the sum of field strengths. This process has to be carried out only once and thus can be easily accomplished by numerical means [Bru92].

The other possibility to statistically combine the set of signals is to operate in the regime of log-normal distributions. Even if the joint probability distribution function of two log-normal distributions is log-normal, too, it is in contrast to the Gaussian case, not possible to calculate the mean value and the standard deviation of the resulting log-normal density function by analytical means. Instead, rather complicated numerical calculations have to be carried out. For an application to real world problems the corresponding time consumption is quite large. Some years ago this created real problems with respect to the existing computer performance. Meanwhile, it is no longer a problem. So an exact numerical solution should be always feasible. If performance is still an issue the exact numerical treatment can be substituted by employing an approximation based on bilinear interpolation [Phi95].

The latter improvement has facilitated the application of LNM methods to a large extend. Usually, the first method was utilized for extensive investigations due to the advantages with respect to computing time. However, it has been evident all the time that the method lacks accuracy. So, when compared to complete Monte-Carlo simulations the results obtained by Gaussian distribution based LNM were not very satisfying. This drawback does not exist with the second approach.

4.5.4 Other Methods

All methods presented so far are based on the assumption that by calculating field strength levels and combining them appropriately, it is possible to arrive at some conclusion about the coverage probability. However, strictly speaking already for analogue transmission systems where all these methods described above have their origin this idea is a very strong simplification of reality. For digital systems such as T-DAB and DVB-T this is even more true. The ratio of the wanted and the unwanted signal levels is certainly an important criterion since without sufficient field strength at the point of reception no receiver will be able to demodulate the transmitted information. Unfortunately, the inverse conclusion that enough field strength will cure any reception problems is only conditionally valid.

The adequate quantity that allows to assess the reception quality in most cases is the bit error ratio. T-DAB and DVB-T are equipped with very efficient error protection mechanisms as described in Sections 2.2 and 2.5. This means that also in areas where wave propagation models predict only a very low field strength an immaculate reception quality can be found. On the other hand, multipath effects in particular, when guard interval violations are encountered, can lead to completely failed reception.

The correct way to derive a criterion that would allow the assessment of the transmission would try to establish a link between the radio channel characteristics and the bit error ratio. First steps in that direction have been made (see e.g. [Beu98a] and [Kue98]). A full description of the situation in all aspects is still to be expected.

Chapter 5

Frequency Planning Basics

Frequency planning is the process of assigning frequencies or channels selected from a set of available frequencies to a set of transmitters or geographical areas. However, the assignment of frequencies has to be accomplished under the condition to meet two fundamental constraints. First, each transmitter should be able to provide a defined quality of service within a specified area and second, the frequency assignments have to be so carried out such that the transmitters do not cause harmful interference to each other. Even though this is easily said, the process is very intricate and calls for appropriate methods to cope with it.

Basically, the process of compiling a frequency plan comprises three fundamental steps. First, a set of requirements has to be defined onto which frequencies or channels selected from an available part of the electromagnetic spectrum are assigned. The technical characteristics of these requirements need to be specified in detail in order to assess the mutual compatibility of each pair of requirements. This constitutes the second step of the frequency planning process resulting in the information which requirements can share and which cannot share a frequency. In the latter case, it is helpful to obtain some kind of measure for the interference to be expected if the same frequency would be used anyway. Finally, the results of the compatibility assessment is fed into an appropriate frequency assignment algorithm leading to a frequency plan in the end.

5.1 Assignments and Allotments

In order to broadcast a certain program throughout a larger area, such as a whole country, not just one transmitter but a set of transmitters is

R. Beutler, *Digital Terrestrial Broadcasting Networks*,
DOI 10.1007/978-0-387-09635-3_5, © Springer Science+Business Media, LLC 2008

necessary. Each of these transmitters that are described in terms of a limited set of technical parameters such as ERP and antenna diagram (see Section 4.1) needs to be assigned a frequency or a channel. Transmitters together with their technical characteristics and the assigned frequency constitute entries of agreed frequency plans. Such plan entries are usually called assignments. In order to generate a frequency plan that is based on assignments, the technical characteristics of all transmitters that are to be assigned a frequency need to be specified in detail. This refers in particular to geographical coordinates, ERP, and antenna height and pattern. This information is used explicitly in the process of finding appropriate frequencies for all transmitters.

With the advent of digital terrestrial broadcasting systems based on COFDM technology new ways to build transmitter networks could be employed (see Section 2.1). Instead of using a different frequency for each transmitter the concept of single frequency networks (SFN) has been developed. These consist of a set of transmitters which all broadcast the same content on the same frequency or channel. The special structure of COFDM signals opens this possibility. SFN operation gives the network provider a lot of freedom to design the network, that is to select the number of transmitters and their geographical location. This was the starting point for the development of a new planning concept for frequency planning, which is based on so-called allotments. Defining an allotment means in the first place to specify a certain geographical area like, for example, the whole territory of a country or a part of it throughout which a particular service is provided. This service is broadcast with the help of a single frequency or channel in this area. The usage of the assigned frequency inside the allotment area is only subject to the condition that the field strength values that are generated by a network inside the allotment do not exceed certain limits outside the allotment area.

Within the ITU framework both assignments and allotments are well-known concepts. The RR [ITU04] give a very general definition that can be applied to other distribution paths like satellite broadcasting as well. They read for an assignment:

> *Assignment (of a radio frequency or radio frequency channel): Authorization given by an administration for a radio station to use a radio frequency or radio frequency channel under specified conditions;*

and for an allotment:

> *Allotment (of a radio frequency or radio frequency channel):*
> *Entry of a designated frequency channel in an agreed plan,*
> *adopted by a competent conference, for use by one or more*
> *administrations for a terrestrial or space radiocommunication*
> *service in one or more identified countries or geographical*
> *areas and under specified conditions.*

The definition of an allotment is more general and abstract. However, the major difference between these two concepts concerns the implications of their usage in the process of frequency plan generation. An assignment corresponds to a transmitter that is specified in all details. When entered into a frequency plan it gives the right to implement exactly these technical characteristics. This governs the level of interference the station is allowed to produce that might impact the reception of signals from other transmitters. It does not say anything about the service area of the assignment nor its protection.

This is slightly different in the case of an allotment that is primarily defined by its associated geographical area. This area is interpreted as the area that is to be served by any network implementing the allotment. In order to be able to produce a frequency plan containing allotments, some measure of the interference a transmitter network implementing the allotment will produce is necessary. Therefore, reference planning configurations (RPC) and reference networks (RN) have been developed to define some kind of placeholder that can be used for frequency planning purposes (see Sections 5.4 and 5.3). The level of interference emanating from an network, making use of an allotment is limited by the characteristics of the associated RPC and RN. This is basically the same philosophy as in the case of an assignment. But, the existence of an associated geographical area gives rise to the possibility to limit interference from outside the allotment area. Hence, an allotment entry in a frequency plan gives the right to produce a certain level of interference to other networks and—very important—implicitly protects the envisaged services area from other networks.

Both assignments and allotments can be used as basic objects for the generation of a frequency plan. Even though their description in terms of technical parameters might be quite different, they can be treated in the same way from a mathematical point of view. Assignment planning assumes that all characteristics of a transmitter are known. As a consequence, a service area can be deduced from these details. This can be approximated by a corresponding polygon and subsequently treated as an ordinary allotment in an allotment planning process.

5.2   Definition of the Service Area for Assignments and Allotments

In Section 4.5 the basic definition of served, interfered, and not served pixels, that is points of reception has been given. Naturally, the set of all served pixels defines the services area. This applies for both assignments and allotments. However, there is a fundamental difference in the way how the service area is connected to the characteristics of an assignment or an allotment. As mentioned in the previous section, an assignment plan entry only gives the right to produce a certain level of interference, but there is no protection of any service area of the assignment. Rather, its service area is the result of the interaction between the signal from the corresponding transmitter and its spectral environment, that is distant transmitters using the same or adjacent frequencies. This is quite contrary to an allotment where the service area is defined from the very start and as part of the concept is guaranteed protection against interference from other transmitters during the generation of a frequency plan.

The specification of a service area by a national authority is usually governed by national media politics and economics. It might be important to serve a whole country at every single point or just parts of it. In other cases, only big cities or major traffic routes might need to be provided with a corresponding service. As a consequence, service areas cannot be given in terms of simple geometrical shapes like circles or quadrilaterals. A service area corresponding to a particular political or cultural requirement need not necessarily be simply connected in a mathematical sense. There might be holes in the middle of the service area or it might consist of several not connected and isolated pieces.

The standard way of describing such geometrical two-dimensional objects is to employ polygons as an approximation to the real shape. This is simple enough in terms of carrying out calculations on these polygons and nevertheless leaves enough freedom to increase the accuracy of the approximation simply by increasing the number of vertices of the polygons. Figure 5.1 shows a simple example of conceivable service areas from Germany.

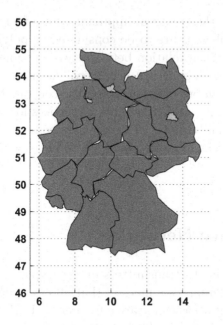

Figure 5.1: Example for a set of allotment areas that are given in terms of polygons.

It is straightforward to approximate service areas by polygons whose vertices are given in terms of geographical coordinates. However, this requires to specify the coordinate system to be used at the same time. One of the basic mathematical operations to be applied over and over again in the course of frequency plan generation is to calculate the distance between two geographical points such as, for example, the transmitter and the receiver location. For calculations on an international scale the earth's curvature cannot be neglected. In most cases, it is, however, sufficient to approximate the earth by a sphere. More elaborated analyses that assume the earth being an ellipsoid or a geoid are usually not necessary. The error which is made by relying on the spherical approximation is negligible in comparison with other error sources. Already the approximation of service areas by polygons introduces variations with respect to the actual location of the boundaries in the order of 10 or 20 km. The differences between distances calculated on the basis of spherical or ellipsoidal approximations are by far smaller.

When applying the spherical approximation of the earth, it is clear that the service polygons are to be considered as polygons on a sphere. Thus, every geometrical relation between two polygons has to be investigated on the basis of spherical geometry. As usual, the polygons are specified by their number of vertices and the coordinates of each vertex. Usually, the maximum allowed number of vertices for a single polygon is defined. For national borders, 99 vertices are common, whereas regional or local coverage areas in most cases have to be approximated with only 36 vertices.

5.3   Reference Networks

Assigning frequencies to transmitters or geographical areas during the frequency plan presupposes a method that allows to decide if two transmitters or allotment areas can share a frequency. In the case of transmitters this is straightforward. Their technical characteristics are known, that is the location of the transmitter site, ERPs, antenna diagrams, and antenna heights above ground. Therefore, an appropriate wave propagation model can be employed to calculate the field strength that is produced by a transmitter at any arbitrary point. This allows to assess the coverage status of any point of reception by comparing with the minimum required field strength. If all contributions from other interfering transmitters at that point are known, it can be checked if the required protection ratio between the useful signal and the sum of all interfering signals is exceeded or not (see Section 4.5).

The situation is quite different for allotments. In the first place they are defined by providing the vertices of a polygon representing the envisaged service area. No transmitter characteristics are associated with that area that could be used to carry out some field strength calculations. As mentioned earlier, the philosophy of the allotment concept is not to make use of any detailed transmitter data at the stage of frequency planning. Rather, a frequency is assigned to the allotment area that can be used under defined conditions as described in the ITU definition of an allotment in the previous section. But then it is necessary to specify what these conditions are. To this end, the concept of RN has been introduced [OLe98]. They are to be used as placeholders that define an interference envelope or umbrella of an allotment. Any real network implementation within the allotment area has to stay below this envelope at any point outside the allotment area.

Allotment planning for digital terrestrial broadcasting systems is guided by the possibility to build SFNs. In contrast to networks based on individual transmitters using distinct frequencies, SFN technology gives the network provider a lot of freedom when designing the network. The number and location of transmitters can be adjusted to different coverage targets. Apart from economic aspects the configuration of the network is governed by the need to provide a certain field strength throughout the envisaged service area in order to offer a particular service quality. This has a direct influence on the inter-transmitter distance in a SFN. Depending on the service quality distances of 10–60 km are typical values. Based on this observation an allotment is associated with a virtual network, namely the reference network, that consists of a small number of transmitters arranged in a symmetrical manner. Each of the employed transmitters has precisely defined technical characteristics. Figures 5.2 and 5.3 show two different reference networks, namely an open and a closed network.

RNs of the type shown in Figures 5.2 and 5.3 have been used for the CEPT T-DAB planning conference at Wiesbaden for the VHF band and L-Band [CEP95]. Both reference networks consist of seven transmitters. They are laid out in the form of a hexagon with one transmitter located in the center. The other six have their positions at the vertices of the hexagon. According to the different propagation conditions in VHF and L-Band different network parameters were chosen. The distance between two transmitters in the VHF reference network of Figure 5.2 is 60 km,

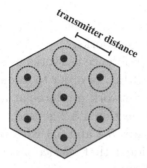

Figure 5.2: Open reference network. The dashed lines sketch the antenna diagrams of the individual transmitters.

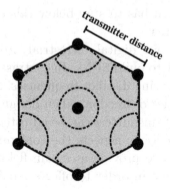

Figure 5.3: Closed reference network. The dashed lines sketch the
antenna diagrams of the individual transmitters.

whereas in L-Band only 15 km were decided to be appropriate. In both
cases, it is assumed that all transmitters have an effective height of 150
m for all directions. The power of the central transmitter is smaller than
that of the other transmitters.

The designations "open" and "closed" refer to the antenna diagrams
that are used. In an open network, all transmitters are equipped with
omnidirectional antennas while in a closed network the transmitters at
the boundary of the envisaged service area have directional antennas having their maximum lobes directed towards the center. Forward–backward
ratio of the antenna diagrams of 12 dB with an aperture angle of 120°
are often assumed.

The gray shaded areas in Figure 5.2 and 5.3 represent the calculated
service area of the reference network without any external interfering
transmitter taken into account. Table 5.1 summarizes the relevant parameters of both reference networks.

Calculations based on reference networks usually employ ITU-R Recommendation P.1546 [ITU01a]. In order to calculate the interfering field
strength that a reference network produces at a given point, the power
sum method discussed in Section 4.5.2 is used.

When comparing Figures 5.2 and 5.3 with the polygons depicted in
Figure 5.1 it becomes evident that service areas of RN do not coincide
with the real shapes of allotment areas in general. This fact created and
still is creating some confusion among people involved in the business
of frequency management and network planning. In particular, the idea

Table 5.1: Parameters of reference networks for VHF and L-Band.

	VHF	L-Band
Type of network	Closed	Open
Inter-transmitter distance	60 km	15 km
Effective heights	150 m	150 m
Power of central transmitter	100 W	500 W
Antenna diagram of central transmitter	Omnidirectional	Omnidirectional
Power of transmitter at periphery	1 kW	1 kW
Antenna diagram at periphery	Directed towards center	Omnidirectional
Diameter of coverage area	<= 120 km	<= 60 km

was and unfortunately is still widespread that inside an allotment area a network is to be implemented that corresponds to the geometry of the RN, for example, seven transmitters laid out as a hexagon. Clearly, this is not true. The reference networks only serve the purpose of estimating the outgoing interference with the help of a hypothetical network whose size is adopted to the frequency range under consideration.

Still the question is how a RN can be used in connection with an arbitrarily shaped allotment area as shown in Figure 5.1. To this end, it is necessary to define corresponding rules. The idea is to position and orient the RN at the border of the allotment area and then calculate the total field strength at a given point of reception. Figure 5.4 sketches the situation.

For each calculation point under consideration outside the allotment area, this calculation is carried out at each vertex of the allotment. The

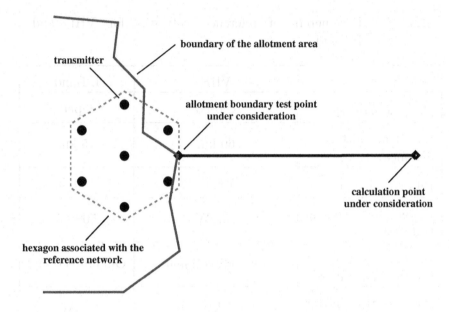

Figure 5.4: Positioning and orientation of reference network with respect to the boundary of an allotment area.

maximum value obtained from all these calculations is then defined as the interfering field strength value any network implementation of the allotment must not exceed at that point.

The figures given in Table 5.1 have to be considered as defining only a very rough estimate of the behavior of a real network. This becomes clear when looking at the concept of RN in general. Neither the network configuration nor the wave propagation model represents the real world with a high accuracy. Reference networks do not take into account any topography. So, they are applied in Netherlands and in the Alps without any modification. It is obvious that this cannot be completely correct. In order to better adapt the concept of RN to the real world it became evident that further types of RN were needed. They have been developed in recent years. This is particularly important in connection with small allotment areas that are supposed to serve only cities or small parts of a country. More details on reference networks and their technical characteristics can be found in [Bru05].

5.4 Reference Planning Configurations

The preparation of input requirements in terms of defining the characteristics of transmitters or allotment areas is not sufficient for a successful
compilation of a frequency plan. Actually, it is necessary to specify
which type of reception is envisaged for each assignment or allotment.
For terrestrial broadcasting there are four types that are distinguished.
They need to be dealt with independently. These are fixed, portable,
handheld, and mobile reception. Portable is further distinguished between portable outdoor and portable indoor reception. Apart from fixed
reception, all reception modes presume the usage of a simple rod antenna. The fixed case is the traditional way of receiving radio signals
where a highly directional antenna mounted at the roof of a building
is employed. Portable reception refers to reception with a typical radio
or TV set that can be moved from one location to another. However,
it is used in stationary mode then. In contrast, mobile indeed refers to
reception in cars or public transport systems whil being on the move.
Handheld is a special reception mode that is to reflect the way mobile
telephones are used, that is with special small-sized handheld devices.
In addition, handheld can be either under portable or mobile reception
conditions.

It is evident that all these different reception modes require different
minimum field strength values to be provided at the point of reception. This means, in particular, at the appropriate height above ground.
For fixed reception usually an receiving antenna height of 10 m is assumed. Portable outdoor and mobile reception correspond to a standard
receiving antenna height of 1.5 m. Portable indoor can be at any height
between 10 m and 1.5 m. From 10 m to 1.5 m, a decline of the field
strength in the order of 10–16 dB can be observed depending on the
wave length. Furthermore, the antenna gain connected to the usage of
directional antennas as in the case of fixed reception has an important
influence on the minimum field strength as well. For indoor reception, a
penetration loss caused by walls and windows needs to be considered in
order to make allowance for field strength attenuations.

Usually, the minimum field strength values that are circulated refer
to 10 m above ground even for those reception modes where reception at
1.5 m is implied. This is accounted for by correcting the minimum field
strength at 10 m by a so-called height loss. Positive antenna gains are

taking into account by subtracting a corresponding factor. Effectively, the minimum field strength given for planing purposes is to be interpreted as the field strength that is to be provided at 10 m height in order to assure good reception under the conditions concerned, that is at 1.5 m or inside a building. A description of the method how the minimum field strength can be calculated systematically can be found in [EBU01].

So far, only the four fundamental reception modes have been addressed. However, they can be combined in principle with any chosen system variant of a given broadcasting system as described in Sections 2.2 and 2.5 for T-DAB and DVB-T. Each of these corresponds to different protection ratio values, depending on the employed modulation and data rate. Table 2.3 shows only a small number of the conceivable system variants for DVB-T. There exist a lot more [Rei01], some of which, however, might be only of academic interest. But in principle, each of these variants leads to different minimum field strengths according to [EBU01].

From a frequency planning point of view such a large number of different possibilities is difficult to deal with. It only increases the complexity of the frequency assignment process in the first place. In order to simplify frequency planning for digital terrestrial broadcasting the so-called RPCs have been developed for the preparation of the GE06 Plan [ITU06] (see Chapter 8 for more details on the GE06 Plan). The idea is to condense the vast number of different options to only a few representative possibilities that are explicitly taken into account. The parameters of the RPCs have to be chosen such that the most important combinations of reception mode and system variant can be approximated by one of the RPCs. In the case of the GE06 Plan, which is a frequency plan for T-DAB and DVB-T five RPCs have been defined, two for T-DAB and three for DVB-T. They are meant to represent portable outdoor and mobile reception in the case of T-DAB and fixed, portable outdoor and portable indoor reception for DVB-T, respectively. Tables 5.2 and 5.3 lists the most important planning parameters of these RPCs. The minimum field strength values have to be interpreted as those values that need to be provided at a height of 10 m above ground in order to guarantee sufficient field strength under the conditions envisaged. For portable outdoor and mobile reception, this refers to a receiver at 1.5 m above ground, while fixed reception assumes a directional receiving antenna at 10 m. A more detailed description of RPCs can be found in [Bru05].

Table 5.2: RPCs for DVB-T used at the RRC-06.

	RPC1	RPC2	RPC3
System	DVB-T	DVB-T	DVB-T
Location probability	95%	95%	95%
Protection ratio	21 dB	19 dB	17 dB
Minimum field strength at 200 MHz	50 dBµV/m	67 dBµV/m	76 dBµV/m
Minimum field strength at 650 MHz	56 dBµV/m	78 dBµV/m	88 dBµV/m

Table 5.3: RPCs for T-DAB used at the RRC-06.

	RPC4	RPC5
System	T-DAB	T-DAB
Location probability	99%	95%
Protection ratio	15 dB	15 dB
Minimum field strength at 200 MHz	60 dBµV/m	66 dBµV/m

5.5   Co-channel Reuse Distances

Once having attributed both RNs and RPCs to a set of input requirements, it is possible to calculate the interference a typical network implementation within a given allotment area will produce at particularly
chosen points of reception. On the other hand, by specifying the RPC the
protection an assignment or an allotment needs to be granted is known
explicitly. Both aspects determine the distance that two allotments need
to be separated to allow them to make use of the same frequency or channel. This particular distance is called co-channel reuse distance.

The reuse distance depends on several parameters, in the first line on the chosen RN and RPC. But, also the frequency plays a crucial role. Electromagnetic waves in the VHF band are significantly more far reaching than in L-Band, for example. However, this is only true as long as propagation above land level is concerned. Above sea level the situation might be different.

Co-channel reuse distance is a theoretical concept that rests on RNs and RPCs. Certain combinations of RN and RPC can be used to carry out calculations in order to determine the reuse distance. Since RNs are involved all calculations are based on ITU-R Recommendation P.1546 [ITU01a]. This guarantees a globally accepted basis for the calculations even though real-world reuse distances between co-channelled networks might be totally different from the theoretical values. It is evident that any discrepancies are due to topographical and morphological conditions that are not properly captured by the wave propagation model as defined by ITU-R recommendation P.1546.

From a theoretical point of view the reuse distance is defined as that distance between the borders of the service areas of two RNs when both networks do not impose an unacceptable level of interference onto the other. This is visualized in Figures 5.5 and 5.6.

Even though such a definition can be intuitively understood care has to be taken about what is actually meant by that. It is evident that if both networks are separated by a large distance their mutual interference is small. The closer they get, however, the larger the interfering field strength from one network into the other becomes. The power budget of

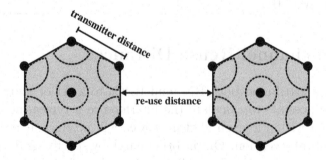

Figure 5.5: Definition of the co-channel reuse distance for a closed reference network.The dashed lines sketch the antenna diagrams of the individual transmitters.

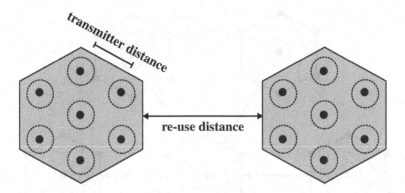

Figure 5.6: Definition of the co-channel reuse distance for an open reference network. The dashed lines sketch the antenna diagrams of the individual transmitters.

the RN is such that all points inside the gray area are served as long as no interference is present. However, when two RNs are shifted towards each other, there is a distance where the first pixel inside the service area changes its status from served to interfered. The corresponding separation distance between the borders of the service areas of the two allotments is then called the co-channel reuse distance.

Typically, the reuse distance is calculated by employing a symmetrical hexagonal geometry where one reference network is surrounded by six reference networks. Their separation distance is systematically changed in order to determine the distance where the first pixel inside the service area of the central reference networks switches from "served" to "interfered." Figure 5.7 sketches the geometrical layout.

The calculation method visualized in Figure 5.7 implies a very particular interpretation of the term co-channel reuse distance. It is assumed that the envisaged service area has to be fully protected, that it is not accepted that a single point inside the service area is affected by the interference of other networks using the same frequency. In other words, the available spectrum is used in a way to protect coverage targets of individual networks. Co-channelled networks need to be separated such that the intended service area is not affected at all. Therefore, depending on the RNs and RPCs chosen the spectrum demand in order to satisfy a given set of requirements in a frequency planning process can be rather large.

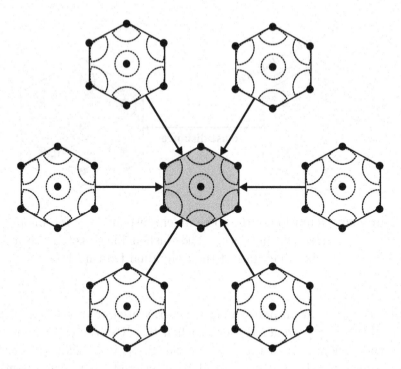

Figure 5.7: Hexagonal layout to calculate the reuse distance.

Another way to calculate the reuse distance is oriented more towards an optimal spectrum usage on a global scale. To this end, the reuse distance can be defined by employing an infinite number of reference networks located at the vertices of a hexagonal grid in an infinite plane. Then, all pixels in the plane are assessed in terms of being served, interfered, or not served. This is carried out for different values of the distance between the borders of the service areas of the reference networks. The maximum number of served pixels encountered defines the co-channel reuse distance. Clearly, from a practical point of view all calculations need to be restricted to areas that are large enough to neglect boundary effects, but nevertheless are large enough to still be computationally feasible. The resulting reuse distance can be interpreted as an interference limited reuse distance, that is co-channelled networks are positioned at distances between such that their individual service area is determined by the interference environment rather than defining an area to be protected. In contrast to the first calculation method, this defines the reuse distance in a way to obtain an optimal usage of a given frequency in global terms, that is it refers to most efficient usage of spectrum.

As mentioned above, the reuse distance depends both on the frequency employed and on the propagation conditions. As the model of ITU-R Recommendation P.1546 is employed for the calculation topography and morphology is not explicitly taken into account. However, for the application of ITU-R Recommendation P.1546 it is important if propagation above land or water is investigated. Water is further distinguished in terms of cold, warm, or even hot sea. The propagation conditions differ dramatically between land and sea propagation. This can be easily seen when looking at the co-channel reuse distances that have been calculated from the RNs of the T-DAB planning conference held at Wiesbaden in Germany in 1995 [CEP95]. Table 5.4 contains the values for VHF and L-Band. It should be noted that the propagation above sea level is more far reaching in L-Band than in VHF leading to significantly increased reuse distances than above land. This is remarkable because above land a larger reuse distance is required for VHF than in L-Band.

Table 5.4: Co-channel reuse distances as calculated from the reference networks of the T-DAB planning conference held at Wiesbaden, Germany in 1995 for VHF and L-Band.

Reuse distance above	VHF	L-Band
Land	81 km	61 km
Cold sea	142 km	348 km
Warm sea	173 km	485 km

5.6 Frequency Plan Generation

In general, the generation of a frequency plan consists of two steps. Once a set of requirements has been identified for which appropriate frequencies or channels should be found, the mutual interference potential between any pair of requirements has to be assessed. This process is called compatibility analysis and allows to decide if two requirements are allowed to share a frequency or not. Based on this information the second step of the frequency plan generation, that is the actual allocation of frequencies or channels can be carried out. This step is usually referred to as plan synthesis.

5.6.1   Compatibility Analysis

The compatibility analysis requires a full description of all technical characteristics of the requirements that are needed in order to determine the level of interference to be expected from one requirement onto another, respectively. In the cases of assignments, this refers to geographical location, ERP, antenna height above ground, and the antenna pattern. If frequencies are to be allocated onto allotments, the vertices of the allotment area as well as RN and RPC have to be specified. Together with an appropriate wave propagation model, this is sufficient to evaluate the level of interference that can be expected at a given geographical point. The result of the compatibility analysis allows to uniquely identify those requirements that can share a channel.

Basically, there are two approaches that can be used for the compatibility analysis. First, a detailed analysis can be carried out on the basis of explicitly calculating field strength values at, for example, the boundaries of allotment areas. Second, the compatibility analysis can be based on a comparison between so-called effective distances between services areas associated with requirements and relevant reuse distances for co-channel usage.

5.6.1.1   Field Strength Based Compatibility Analysis

The field strength based compatibility analysis has been employed in all major planning conferences held so far for digital terrestrial broadcasting systems. This refers to the CEPT planning conference for T-DAB in Wiesbaden, Germany in 1995 [CEP95], the CEPT planning conference for T-DAB in L-Band held in Maastricht, Netherlands in 2002 [CEP02], and finally the Regional Radiocommunication Conference of the ITU which took place in Geneva in 2006 [ITU06].

Basing the assessment of the mutual compatibility of requirements on the calculation of field strength values to be expected at certain geographical points is certainly the more accurate of the two methods. In particular, it allows to investigate the mutual compatibility between very different types of systems and services, provided that the mutual protection criteria are known in terms of corresponding technical parameters such as minimum field strength and protection ratios. The drawback of the method is that it requires a large computational effort. In case of a big planning conference with a corresponding large number of input requirements to the planning process, this might lead to huge computation times.

The application of a field strength based compatibility analysis to a set of assignments and allotments requires to carefully define how the interference produced by an assignment or an allotment is to be calculated at a given geographical point. For an assignment this is straightforward. It basically represents a transmitter site, and therefore, its technical characteristics can be used when applying an appropriate wave propagation model.

The situation is a bit more complicated for an allotment requirement. Basically, it is defined in terms of its allotment area given as a set of vertices and the chosen RN and RPC. In the context of a compatibility analysis the RPC represents the protection requirements, whereas the RN determines the interference produced by the allotment. Assessing the mutual impact of the two allotments, all vertices of one allotment are checked against all vertices of the other allotment.

As already mentioned in Section 5.3, the calculation of a field strength at a given point requires to specify how the RN is positioned in relation to the geometry of the allotment area. Figure 5.4 sketches the standard layout. For a given geographical point where the interfering field strength is to be determined the RN is positioned at each of the vertices of the allotment area according to the Figure 5.4 and the field strength is calculated at the given point of interest. The largest value found is then used as the field strength value of the RN at that point. Figure 5.8 illustrates the procedure to be followed.

Once the field strength imposed by one allotment on the boundary of another allotment has been determined, it can be compared to the maximum allowed value. The maximum allowed field interfering field strength corresponds to the minimum field strength needed to assure reception subtracted by the relevant protection ratio. If the actual field strength is less than that the two allotments can share a frequency, otherwise they are incompatible and cannot be assigned the same channel during the frequency plan generation process.

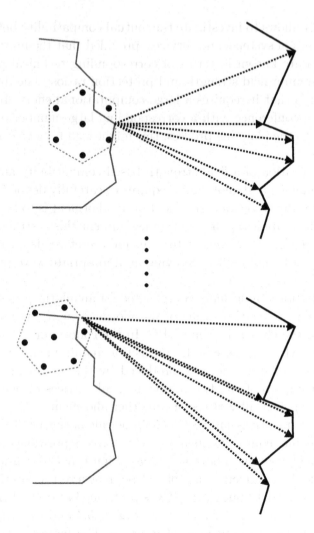

Figure 5.8: Positioning of reference network along the border of
an allotment area in order to calculate the interfering
field strength at the vertices of another allotment.

If assignments are involved one additional issue has to be addressed.
The calculation of the interference produced by the assignment is evaluated on the basis of its technical characteristics. However, an assignment
is not associated with a service area by definition. The service area of an
assignment can only be determined once all co-channel assignments and

allotments are known. Then for each pixel throughout a given area the services quality can be assessed. The set of those pixels labeled "served" will then constitute the service area of the assignment. In other words, the actual service area of an assignment is only know after the frequency plan has been established.

In order to have some indication that two assignments or an assignment and an allotment can share a channel it is, therefore, necessary to make further assumptions on the compatibility analysis about the service area of the assignment to be protected. A quite simple approach, also employed during the Regional Radiocommunication Conference 2006 (RRC-06) when establishing the GE06 Agreement [ITU06] (more details on the RRC-06 can be found in Chapter 8), is to calculate the contour on which the field strength of the assignment has dropped to the required minimum field plus a correction factor used to account for the impact of interference from other networks. In GE06 a value of 3 dB has been adopted thereto. This contour is then dealt with in the same way as an allotment boundary is, that is its vertices are used as those points where the interfering field strength from other assignments or allotments is evaluated.

5.6.1.2   Effective Distance Based Compatibility Analysis

Figure 5.8 indicates that a field strength based approach to compatibility analysis can be rather time-consuming. This is due to the fact that for the assessment of a pair of allotments the RN has to be positioned at each vertex of the interfering allotment in order to evaluate the impact on one vertex of the affected allotment. And then this has to be repeated for each vertex of the affected allotment area. Depending on the conditions under which a frequency plan is to be established, other approaches might be more favorable.

About two years before the RRC-06 was conducted to establish a frequency plan for digital terrestrial television and sound broadcasting in Europe, Africa, and parts of Asia [ITU06] administrations started to talk to each other in order to coordinate the spectrum usage along their common borders. To this end, several studies were carried out, to first of all evaluate the amount of spectrum that might be needed in order to accommodate all expected requirements for T-DAB and DVB-T. Moreover, in bi- and multilateral negotiations frequency plans have been calculated on a regional level to see how the available spectrum could be distributed among neighboring countries.

For such investigations it is not necessary to take into account all available technical details. Rather, due to the iterative character of bi- and multilateral negotiations it is better to have a quick but still reliable and controllable method that allows to include all relevant features on a more qualitative level. Clearly, such an approximation has to be agreed by all participating parties as a mean to base negotiations on.

Having sampled a set of requirements for which frequencies out of a given limit part of the electromagnetic spectrum are to be allocated the crucial question that has to be answered is whether two requirements can share a channel or not. As discussed above, the calculation of field strength values constitutes a precise tool to find answers to this question. However, the concept of reuse distances can be used as well. It reflects the fact that the electromagnetic field strength decreases with increasing distance from a transmitter. Beyond a certain distance—the reuse distance—a frequency can be reused again without causing too much interference or being too much interfered by other networks. Consequently, the distance between service areas can be used as a measure to decide about potential spectrum reuse. Such an approach is less technical and can be understood intuitively.

Basically, checking the mutual interference potential between two service areas thus means to calculate the minimal geographical distance between them and compare it to the corresponding reuse distance. This has to be carried out for all pairs of allotment areas. To make the discussion straightforward, a very simple example is used in order to explain the context. Five polygons are sufficient for that purpose. They are sketched in Figure 5.9.

From a purely geometrical point of view the minimal distance between two polygons is either a line connecting two vertices or a vertex and an edge. In the latter case the connecting line is perpendicular to the edge. Figure 5.10 illustrates the possibilities.

In order to apply a distance based compatibility approach all mutual distances between the polygons have to be calculated. These distances constitute the elements of a symmetrical distance matrix. Each distance between two polygons is then compared to the corresponding reuse distance. If the distance is larger than the reuse distance the requirements associated to the pair of polygons can share a frequency, otherwise they cannot. This information is typically stored in an additional symmetrical

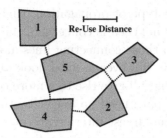

Figure 5.9: Five allotment areas given in terms of simple polygons. The dashed lines indicate those distances between areas that are smaller than the reuse distance.

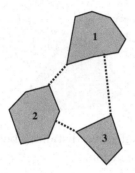

Figure 5.10: Minimal distances between polygons. The minimum separation line connects either two vertices or one vertex and an edge to which it is perpendicular then.

matrix, the so-called adjacency matrix. In Figure 5.9 the reuse distance has been chosen such that an adjacency matrix of the form

$$N = \begin{pmatrix} 0 & 0 & 0 & 0 & 1 \\ 0 & 0 & 1 & 1 & 1 \\ 0 & 1 & 0 & 0 & 1 \\ 0 & 1 & 0 & 0 & 1 \\ 1 & 1 & 1 & 1 & 0 \end{pmatrix} \tag{5.1}$$

results.

As long as only one type of propagation, that is propagation above land or (cold, warm, or hot) sea is concerned, the geographical minimal distances correspond to those connecting lines along which the interference would be maximal. However, a quick look at Table 5.4 shows that mixed propagation paths will give rise to a more complex situation. The geographically shortest path need no longer be the one with the largest interference impact. The largest interference might be imposed along a path that is composed of paths corresponding to different propagation types. Clearly, these different paths will contribute corresponding to their fraction of the total path length. Consequently, instead of a geographical distance between service areas an effective distance between needs to be identified. Figure 5.11 depicts the new situation.

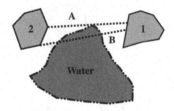

Figure 5.11: Different effective distances between areas for mixed
path propagation.

The two areas 1 and 2 in Figure 5.11 are partially separated by water. Path A is the shortest pure land path. However, it is very likely that the critical path, that is the path whose effective impact is largest, includes a passage over water like, for example, path B. As can be seen from Figure 5.11 such a path must no longer connect two vertices or one vertex and an edge perpendicularly.

If the interaction analysis for two areas is based on field strength curves derived from ITU-R recommendation P.1546 mixed paths can be dealt with according to the interpolation schemes given therein. For the reuse distance approach an adequate formulation does not exist. However, a very simple and evident method that rests on appropriate scaling of distances gives satisfying results.

To explain the idea behind this approach the parameters for the VHF reference network of the CEPT Conference held at Wiesbaden are employed for the time being. According to [CEP95] and Table 5.4 a warm sea path leads to a reuse distance of 173 km instead of 81 km above land.

These different values can be interpreted by saying 173 km path above warm water has the same interference potential as a land path of 81 km. Now, if the distance between two polygons is composed by a fraction of x km above land, y km above cold sea, and z km above warm sea then the combined interference potential is assumed to be equivalent to a situation when the two areas would be separated by a distance

$$d = x + y * \frac{81}{142} + z * \frac{81}{173} \qquad (5.2)$$

above land only. The quantity d is called the effective distance between the two areas. Then, the decision whether the two areas could use the same frequency or not is to be taken by comparing the effective distance d with the reuse distance for land propagation. Naturally, the same method can be applied for any spectral range and any type of reference network as long as the reuse distances for land, cold sea, and warm sea are known.

It has to be noted that the concept of the reuse distances that rests on the wave propagation model according to the recommendation [ITU01a] does not take into account any topographic conditions between the two areas under consideration. Clearly, the adherence of the reuse distance criterion for co-channel usage becomes less critical in reality if both areas are geographically decoupled. To give an example, the frequency plan assembled at Wiesbaden for T-DAB contains polygons on the northern and the southern side of the Alps, whose edges are separated only by a few kilometers. But, due to the high mountains in between they can be considered decoupled without any problems and therefore share the same T-DAB block.

Applied to the method proposed in terms of Equation (5.2) this means that additional information has to be employed when calculating the effective distance. The detection of topographic shielding between two allotment areas could be taken into account. The simplest approach could be that under such circumstances the effective distance between two requirements could be artificially increased to a value larger than the corresponding reuse distance by definition. This would allow a frequency assignment algorithm to allocate the same frequency or channel to the two requirements concerned.

5.6.2 Plan Synthesis

The compatibility analysis provides an adjacency matrix containing the detailed information about any possible co-channel usage of requirements. In order to allocate frequencies onto the requirements appropriate mathematical algorithms are utilized. Basically, two different types of plan synthesis tasks can be distinguished in practice. Since electromagnetic spectrum has become a precious resource it is important to find the minimum number of frequencies or channels that are necessary in order to satisfy a given set of requirements such that all constraints imposed by the adjacency matrix are met. This is usually referred to as a spectrum demand study.

On the other hand, quite often there is only a limited amount of spectrum available for the introduction of a new broadcasting service. It is evident that under such conditions not all constraints relating to a prohibition of co-channel usage between requirements can be met. There will be a certain number of problematic cases where co-channels have to be assigned to requirements that are formally not meant to share a channel. Then, the task is to distribute the available spectrum in a way to minimize the level of interference. Such type of problems are usually called constrained frequency assignment problems.

Form a mathematical point of view, both tasks represent combinatorial problems to be solved. At first glance, at least the spectrum demand problem seems to be an easy task. However, it turns out that in fact the underlying mathematical problem is extremely complex. A short example might give some insight. To this end, it shall be assumed that there are 20 allotments that are to be assigned frequencies. Furthermore, it is supposed that one of ten different frequencies can be assigned to each of the allotments. Under these circumstances there are 10^{20} different possibilities to assign a frequency to each of the polygons. The task is to find an assignment of frequencies onto the allotments that is compatible with the adjacency relations and make use of a minimum of the 10 available channels.

Basically, the only way to tackle such a combinatorial problem is by means of computers. In principle, it can be seen as a trial and error task that means frequencies are allocated to the requirements and then this allocation is assessed whether the adjacency relations are satisfied and if the number of required channels is minimal. How long this takes, clearly depends on the available computing power. One microsecond might be a reasonable value for both the allocation of frequencies and the subsequent evaluation.

A simple but straightforward way to find the optimal solution to this frequency assignment problem is to simply check each of the possible solutions. One starts with an arbitrary frequency assignment and keeps the solution as the best solution found so far. Then, the next one is checked and if it leads to a better result it is stored as the current best solution. This is repeated until all possible way to assign the 10 channel onto the 20 requirements have been evaluated. Such an algorithm is usually referred to as a brute-force approach.

In the case of the simple example at hand, this method would require a time $t = 10^{20} \times 10^{-6}$s $= 10^{14}$s which corresponds to 2.78×10^{10} hours or 1.16×10^9 days or 3.17×10^6 years. In other words, the computer would need to run for more than three million years. Even if a supercomputer would be available that could reduce the computing time of one check to just one nanosecond, still the total time would exceed 3000 years. This simple result proves very clearly that any brute-force approach like the one described here is completely useless in practice. This becomes even more clear when looking at real-world planning exercises. The figures of 20 allotments and 10 available channels might be realistic on a national scale for some countries. On a European level, however, hundreds or thousands of allotments are usually to be taken into account for a planning exercise.

Seen from a mathematical perspective, frequency assignment falls into the class of NP-hard mathematical problems. This means that the computing time to find the optimal solution grows faster than exponentially with respect to the characteristic system size. For a very small number of coverage areas it is possible to calculate the optimal solution by a simple trial and error strategy. However, as can be seen from the example given above already for 20 areas this is no longer feasible.

5.7 Planning Methods and Algorithms

Frequency assignment calls for sophisticated methods to obtain satisfying results due to the complexity of the combinatorial problems that have to be solved. Advanced mathematical algorithms need to be applied in order to obtain frequency plans that use the available spectrum in an efficient way. However, it is also clear that only methods are interesting that deliver results within a reasonable time. Some hours of computation time are not a problem and in the context of a big planning conference like the RRC-06 even some days might be acceptable. Any approach leading to computational times beyond that is worthless from a practical point of view. To date, most planning methods are based on graph theoretical acquirements or employ stochastic optimization algorithms.

5.7.1 Geometrical Allotment Planning

Primarily, generation of a frequency plan is based on the idea that frequencies or channels are allocated to geographical areas. Not only in the case of allotment planning this is obvious, but also for assignments this is true, even though formally the frequency is allocated to a transmitter site rather than an area. However, as discussed in Section 5.6.1.1 every assignment can be associated with a service area in one way or the other. Hence, the allocated frequency is linked to an area as in the case of an allotment as well.

A glance at Figure 5.1 shows that allotment areas can have arbitrary shapes, depending mainly on nontechnical constraints like media-political conditions. Furthermore, the service areas associated with assignments exhibit very different geographical characteristics. In that case, the location of the transmitter site and the surrounding topography are the determining factors for the shape of the service area. Frequency planning needs to take these special layouts into account.

Even though this is evident from a practical point of view, planning methods have been developed and applied in the past, which follow a purely systematic path completely ignoring the geographical boundary conditions. The idea is to distribute channels across a planning area such as, for example, Europe according to a given mathematical scheme. The frequency plan established by the ITU in Stockholm in 1961 for analogue television (ST61) has been based on such an approach [ITU61].

At ST61 only assignments had been planned, but the application of systematic planning methods to allotments is straightforward. If, for example, nation-wide coverages are to be planned, simple geometric areas can be used as basic elements to cover the territory. The whole planning region, for example, Germany or Europe, is then covered by a net consisting of these elements. The only constraint to be met thereto is to employ geometrical elements that allow for a gap free parquet of the plane. Rhombuses or hexagons are very well suited. Circles are usually not used since a gap free two-dimensional coverage of the plane can be generated only if overlapping areas are accepted. Figure 5.12 shows a rhombic net across Germany, whereas in Figure 5.13 a coverage in terms of hexagons is presented.

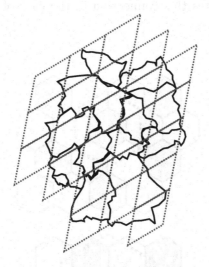

Figure 5.12: Allotment areas given in terms of rhombuses across Germany.

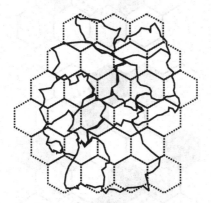

Figure 5.13: Allotment areas given in terms of hexagons across Germany.

Frequencies are assigned to these symmetrical coverage areas by applying a deterministic scheme that respects the reuse distance criterion for co-channel or co-block usage. Depending on the ratio between the characteristic dimensions of the area elements and the reuse distance, different spectrum demands for an entire wide-area coverage results. Figure 5.14 illustrates this connection in the case of planning with the help of hexagons.

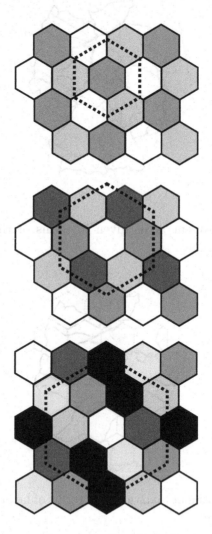

Figure 5.14: Three gap-free parquets on the basis of hexagons. The reuse distance is indicated by the dashed hexagon surrounding the central hexagon. The spectrum demand is 3, 4, and 7 frequencies or channels, respectively.

The spectrum demand for such a hexagonal tiling can be systematically determined when covering the plane with clusters of hexagons without leaving any gaps. Several examples shall illustrate the idea. Different hexagon clusters are considered where the radius of the hexagons, that is the distance from the center to a vertex, is called r. This also corresponds to the length of the edges of each hexagon.

A trivial cluster would be a single hexagon leading to a spectrum demand of just one channel. Certainly, this is of no relevance. The next possibility is to use a cluster of three hexagons to which three different frequencies are assigned. With the help of such a cluster the entire plane can be covered without leaving any gaps by copying and shifting the three-hexagon cluster appropriately. It has to be noted that such a frequency allocation is only valid under the assumption that for the reuse distance R the relation $0 \leq R < r$ holds. The next clusters have the size 4, 7, 9, and so forth. Figure 5.15 shows the first eight hexagon clusters.

The spectrum demand can be described in terms of the so-called rhombic numbers, a sequence of integers that can be derived from the equation

$$\rho_{n,m} = n^2 + nm + m^2 \tag{5.3}$$

with

$$n, m \in \mathbb{Z}. \tag{5.4}$$

The relation (5.3) does not constitute a one-to-one correspondence between a pair of n and m since several combinations of n and m lead to the same value of ρ. However, the rhombic numbers can be ordered in terms of a monotonous series

$$\Omega_k = 0, 1, 3, 4, 7, 9, 12, 13, 16, 19, 21... \tag{5.5}$$

for $k = 1, 2, 3, ...$. Also the maximum and minimum reuse distances giving rise to a certain spectrum demand are given in terms of square roots of rhombic numbers as can be seen from Table 5.5.

Even though the hexagon planning principle is very attractive from a frequency management point of view the question arises how the achieved distribution of frequencies is compatible with national and international media political requirements. A brief look at the Figures 5.12 and 5.13 makes clear that, for example, the boundaries of the German Bundesländer are not respected by the presented layout of rhombuses or hexagons. In case such a principle be applied at a frequency planning

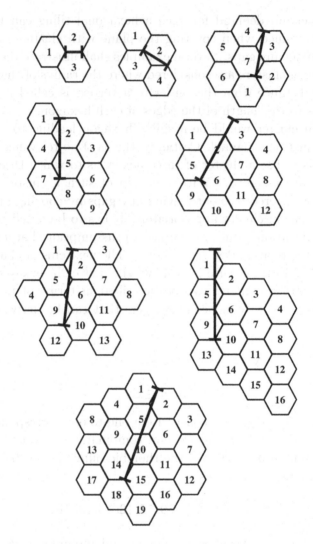

Figure 5.15: The first eight hexagon clusters that allow a gap free coverage of the plan. For each hexagon the corresponding maximum allowed reuse distance is indicated.

conference for the derivation of the frequency plan, it would be obvious that the affected administrations and network providers would need to enter bi- or multilateral coordination negotiations after the conference, to figure out the details of the frequency usage in neighboring countries.

Table 5.5: Spectrum demand of hexagonal tilings of the plane as a function of the reuse distance.

Reuse distance	Number of required channels
$0 \leq R < r$	3
$r \leq R < \sqrt{3}r$	4
$\sqrt{3}r \leq R < \sqrt{7}r$	7
$\sqrt{7}r \leq R < \sqrt{12}r$	9
$\sqrt{12}r \leq R < \sqrt{16}r$	12
$\sqrt{16}r \leq R < \sqrt{19}r$	13
$\sqrt{19}r \leq R < \sqrt{27}r$	16
$\sqrt{27}r \leq R < \sqrt{31}r$	19

Seen from a practical point of view the geometrical allotment planning does not possess a lot of relevance today. However, it has nevertheless great conceptional significance. First of all, geometric allotment planning represents some kind of optimal planning in the sense that the usage of spectrum is optimal. Therefore, it can be employed as a reference in order to assess the efficiency of other planning results.

Moreover, a very simple but, nevertheless, important rule-of-thumb for the generation of allotment areas by administrations and the thereby emerging consequences for the spectrum demand can be derived from its results. The extension of allotment areas in relation to the reuse distance determines the number of required frequencies. Therefore, care has to be taken when defining allotment areas. In order to minimize the spectrum demand the minimal extension of allotment areas should preferably be larger than the relevant reuse distance.

5.7.2 Graph Theory Based Algorithms

Geometrical planning approaches as described in the previous section constitute a theoretical framework resulting in a frequency plan that subsequently has to be mapped onto existing requirements given in terms of allotment areas. In practice, an inverse approach is usually needed. The starting point for the generation of a frequency plan is a set of allotments or assignments specified by all the relevant technical characteristics. At the first step the mutual compatibility relations between all pairs of requirements are derived by calculating a corresponding adjacency matrix. Then, some mathematical algorithm is employed to assign frequencies or channels onto the requirements while respecting the constraints imposed by the adjacency matrix. As discussed in Section 5.6.2 this gives rise to very complex combinatorial problems. In general, it is hopeless to try to find the optimal solution by means of a brute-force approach.

Therefore, the question must be asked if it is really always necessary to find the optimal solution for practical applications. Can it not be satisfactory to get hold of a nearly optimal solution that can be found, however, within a finite and reasonable interval of time? In order to rely on such a work around, it would be very helpful if some criteria would exist giving some indication about how far the accepted nearly optimal and the one and only optimal solution are apart.

In principle, graph theory can deliver both, namely very good approximative solutions and lower bounds for the minimal number of required frequencies. A very good and clear introduction to the field of graph theory can be found in [Jun98]. For a very detailed description of the application of graph theory in particular to problems arising in frequency assignment for T-DAB it is referred to [Gra02]. A very extensive reference for a lot of other different aspects of graph theoretical issues is the Website of the Konrad-Zuse-Zentrum für Informationstechnik in Berlin (ZIB). This Website contains a section, which also provides a very long list of scientific articles connected to graph theory and frequency assignment [ZIB01].

The connection between frequency assignment and graph theory can be established very easily. A situation as depicted in Figure 5.9 can be mapped into a simple graph on the basis of the adjacency matrix (5.1). Figure 5.16 shows the graph corresponding to Figure 5.9. Every polygon shown in Figure 5.9 is represented by a vertex in Figure 5.16. Two areas which according to the adjacency matrix (5.1) are not allowed to use the same frequencies are connected by an edge.

The original task to assign frequencies to areas thus is equivalent to the problem of coloring a graph. This means to assign a color to each of the vertices of the graph such that vertices that are connected by an edge

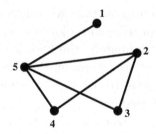

Figure 5.16: Simple graph corresponding to Figure 5.9.

do not get the same color. In practice, the attribute "color" is represented simply by an integer. Clearly, the coloring has to be accomplished under the constraint to use as few colors as possible. The minimum number of required colors for a given graph is called the chromatic number χ of the graph.

What would be the most natural and obvious way to color the graph? Clearly, the most simple thing to do would be to choose an arbitrary vertex and assign a color to it. Then another vertex not yet being dealt with is chosen. It is given a color in accordance with its individual adjacency relations. This means it has to be checked to which other vertices the considered one is linked and which colors have already been given to these. A color not yet used by one of the already treated vertices is then given to the current one. This process is continued until all vertices have been assigned colors. Finally, the number of different colors used is determined. By definition, such a strategy will lead to an admissible solution of the graph coloring problem. However, its drawback is that the quality of the result can be arbitrarily poor in practice.

The quality of the solution can be improved considerably by taking into account the adjacency relations of the vertices in an appropriate way. When looking at the graph of Figure 5.16, it becomes obvious that all vertices have different types of adjacency relations. Vertex 1 only has one single neighbor while vertex number 5 is connected to all other vertices. These differences can be quantified with the help of the so-called vertex degree κ. The degree of a vertex corresponds to the number of edges ending at the vertex under consideration. Coloring methods making explicit use of the vertex degrees of the graph will be presented below in the following subsections in more detail.

Clearly, graphs containing vertices with large degrees very likely will require a large number of different colors to satisfy all adjacency relations appropriately. Unfortunately, the knowledge of all vertex degrees of a graph is not sufficient to determine the chromatic number χ. It is not even sufficient in general to derive a satisfying lower bound for the minimum number of required colors.

This can be understood when looking at Figure 5.16. Vertex 5 has degree $\kappa = 4$. It can be shown, however, that three colors are needed for this simple graph. So obviously, the maximum vertex degree of a graph is not necessarily linked to the chromatic number.

In order to find a reliable bound for the chromatic number, it is necessary to consider subsets of vertices, so-called subgraphs, instead of dealing with individual vertices alone. Moreover, it is crucial to identify complete subgraphs. A complete graph is a graph whose vertices are all mutually connected by edges, that is all vertices of a complete subgraph have the same degree κ. The number of vertices ω of a complete graph and its vertex degrees κ are linked by

$$\omega = \kappa + 1 \ . \tag{5.6}$$

Sometimes a complete subgraph is also called a "clique" and its number of vertices ω is the clique number.

Each graph, which does not consist only of isolated vertices contains complete subgraphs. The reason is that already two connected vertices constitute a trivial complete graph. Figure 5.17 shows a simple example of a graph having several nontrivial complete subgraphs. The vertices {4, 5, 6} or {2, 3, 4, 5}, for example, form complete subgraphs.

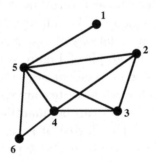

Figure 5.17: Graph containing nontrivial complete subgraphs.

An important parameter connected to the existence of complete subgraphs is the maximum clique number ω_{max}, which is nothing but the number of vertices of the largest complete subgraph of the considered graph. Clearly, the maximum clique number ω_{max} determines the chromatic number to a large extent. Indeed, it can be considered as a lower bound for χ, that is

$$\omega_{max} = \kappa_{max} + 1 \leq \chi, \tag{5.7}$$

where κ_{max} is the corresponding degree of the vertices of the maximum clique.

In most cases, the maximum clique number ω_{max} represents a very good lower bound in the sense that the chromatic number χ is not much larger. The problem, however, is that the determination of the clique number falls into the class of NP-hard problems, too, like the graph coloring itself. Thus, brute-force approaches to find the chromatic number are usually of no help.

In Figure 5.17 the set of vertices $\{2, 3, 4, 5\}$ not only forms a complete subgraph, but at the same time the vertices also form the maximum clique. For this particular graph the chromatic number is 4, which is equal to the maximum clique number. In general, this need not be the case.

For a graph consisting of six vertices it is no problem to determine the maximum clique number even by trial and error. For medium size graphs, that is a number of vertices in the order of 1000, it is possible to exploit an exact algorithm that provides the maximum clique number and the vertices forming the maximum clique within an acceptable period of time [Car90].

5.7.2.1 Sequential Graph Coloring

Vertices having large degrees might lead to a high demand for colors. However, this statement has to be interpreted with care as seen above. Nevertheless, it is reasonable to start the coloring process with those vertices being linked to many others. This is guided by the hope that there is enough freedom for the rest of the vertices to assign a color once the difficult ones are dealt with without increasing the overall consumption of colors.

Indeed, it became evident that such strategies are likely to produce good results. But, it should be born in mind that even if algorithms are utilized, which are based on strict and clear rules, it cannot be guaranteed that the optimal solution will be found this way. Normally, this will not be the case. Hence, all deterministic algorithms, which do not necessarily lead to the determination of the optimal solution like a straightforward trial-and-error approach are called heuristics.

The most simple way to color the apparently critical vertices at the beginning is to establish a sequence of the vertices according to their degrees. Typically, a descending ordering is applied. However, the inverse ordering from small to large degrees can give good results, too. Then, all vertices are sequentially given colors such that each assignment is in accordance with the corresponding adjacency relations.

In terms of mathematical algorithms the attribute "color" of a vertex is usually represented by an integer. So, applying a sequential coloring strategy means, for example, to assign the number zero to the first vertex of the sequence. The next one gets the smallest integer, which is compatible with all its neighboring vertices that already were assigned a number. This is continued until each vertex has been dealt with.

Experience shows that proceeding along a descending order of vertex degrees is more advantageous than an ascending order of vertices. However, one should in any case check both possibilities because it cannot be excluded that another ordering might give a better result. Table 5.6 shows the assignment of colors for the graph depicted in Figure 5.16 obtained by employing a vertex sequence of descending degree order.

Table 5.6: Result of the sequential coloring for the graph shown in Figure 5.16.

Vertex	5	2	4	3	1
Color	0	1	2	2	1

The arrangement of vertices according to descending or ascending vertex degrees obviously is not the only possibility. The so-called "smallest-last" ordering is another very famous and promising way to build a sequence [Mat83]. In contrast to the aforementioned schemes the generation of this sequence is more complex.

The first step is to set up a sequence with descending vertex degree. The vertex at the end of the series, that is the vertex having the smallest degree, is shifted to the last position of an auxiliary sequence. Then this vertex is removed from the original graph together with all the edges ending in that particular vertex. Thereby a new graph is created, which has one vertex less than the original one.

For this new graph once again a vertex ordering based on descending degrees is established. The last vertex of this sequence is shifted to the last but one position of the auxiliary sequence. As before, a new graph is build by removing this vertex from the actual graph, thus giving rise to the second new graph having two vertices less than the original one. This procedure is continued until all positions in the auxiliary sequence are finally filled. The subsequent coloring of the vertices then uses the ordering as given by the auxiliary sequence following the same scheme as employed in the case of Table 5.6.

If several vertices have the same minimal degree, additional criteria have to be applied in order to decide, which vertex is to be removed from the graph. To this end, there are different possibilities. One of the vertices can be chosen by chance or that vertex is removed, which had the smallest degree of all vertices under consideration with respect to the original graph. Figure 5.18 sketches the principle approach of the smallest-last ordering strategy.

Other systematic ordering strategies have been investigated in recent years. The ordering according to the maximum saturation is one of the

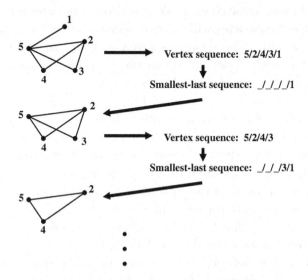

Figure 5.18: Smallest-last ordering of the graph 5.16.

most famous approaches [Bre79]. For further methods it is referred to
[Wei98]. Unfortunately, there is no generally valid rule which suggests
one or the other procedures in a given situation. Therefore, it is advised
to always try several different sequences. This strategy has been followed
at the Wiesbaden Conference, too, where over 1000 different ordering
schemes have been exploited [OLe98].

Furthermore, it has to be emphasized that a graph coloring, which
is carried out on the basis of one of the methods described here implies
that all constraints in terms of adjacency relations are strictly obeyed. In
case two vertices are linked by an edge they are strictly assigned distinct
colors. Speaking in terms of frequency assignment this means the frequency plan derived thereby does not contain any additional interference
beyond the accepted and unavoidable levels.

5.7.2.2 Weighted Graphs

The graphs shown in Figure 5.16 or 5.17 represent very special frequency
assignment problems. Two vertices can either get the same color or not.
Basically, this is a clear and mutually exclusive decision. However, real
frequency assignment problems are usually more complex and cannot
always be mapped to a simple black-or-white decision.

A typical more involved example would be to incorporate constraints
on how far the frequencies which are assigned to two connected vertices
must be separated. The problems given in terms of the Figures 5.16
or 5.17 tacitly imply a separation distance of one channel or block or
whatever frequency unit is actually addressed.

Another task could be to find a frequency assignment for a set of
coverage areas where certain areas using the same frequency would give
rise to a certain penalty to be borne, for example, a level of interference
between them. Then, the task would be to find a distribution of colors
onto the graph such that the total interference produced is minimal.

One way to represent both problems in terms of graphs is to introduce
the concept of weighted graphs. This means that each edge can be provided with a number that has to be interpreted accordingly. Figure 5.19
shows a corresponding extension of the graph 5.16.

The interaction between two vertices in the graph of Figure 5.19
is symmetric. Sometimes it is necessary to introduce an asymmetric
interaction and to operate on the basis of weighted and oriented graphs.

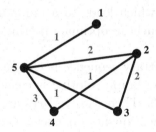

Figure 5.19: Simple weighted graph. The numbers at the edges might represent the number of channels the two frequency assignments must be separated.

Thereby, the complexity of the graph coloring problem is even more increased.

In principle, the same type of approaches to find a coloring as described above can be applied. However, it must be emphasized that the more constraints are to be met the less effective usually such simple and straightforward strategies become. Hence, a lot more effort has to be put into the development of specially adapted coloring methods. Unfortunately, this means they lose their general applicability at the same time.

5.7.3 Frequency Plan Synthesis by Stochastic Optimization

The more constraints have to be met during a graph coloring process the more complicated and complex the problem becomes. Deterministic coloring strategies tend to loose their efficiency under such conditions. Fortunately, there exist other methods that can be used instead. In order to solve highly complex frequency assignment problems stochastic optimization algorithms can be employed. Details concerning the mathematical background can be found in Annex A.

In principle, there are several possibilities for the application of stochastic optimization methods. A first and very straightforward utilization can be identified already in the case of the simple sequential coloring. As has been described above, the ordering of the vertices is the crucial factor that determines the quality of the solution. The so-called Great Deluge algorithm described in detail in Section A.1 could be used to find a sequence of vertices, which leads to a minimum number of required colors. The search for such an optimal ordering is carried out in terms of an iteration.

To this end, an arbitrary sequence is chosen as the starting configuration and the resulting graph coloring is calculated by application of the sequential procedure. Then, two vertices are chosen arbitrarily and their positions within the sequence are exchanged. Again, the colors are assigned. The next step would be to decide, if the second result should be kept as new starting point for the progression of the Great Deluge algorithms or not. By iterating this process a very good solution can be found.

It must be pointed out, however, that such an approach is only reasonable in practice, if the number of vertices is large enough. For a small or moderate number of vertices the variance of the results referring to two different vertex orderings is too small. Most of the sequences will lead to the same number of required colors. So, for the iteration process it is not obvious, which of them is to be chosen as the new starting point.

If in such a situation, the old configuration would always be kept, the algorithm would carry out more or less a local search in the vicinity of the starting configuration. Since this sequence has been chosen by chance it can give rise to a very high color consumption. In the opposite case, when always the new configuration is chosen the algorithm proceeds along an arbitrary path in configuration space. Then the stochastic optimization process bears resemblance to an unrestricted Monto-Carlo simulation.

A further possible application of stochastic optimization to frequency assignment problems is connected to principle issues concerning the applicability of graph theory for that purpose. Situations are quite common, where the available spectrum is so limited that none of the constraints can be satisfied exactly. Also, there might be completely different types of constraints to be obeyed at each vertex.

All these complications are not very easily modeled in terms of graph coloring problems. If a mapping to a corresponding graph coloring problem can be established, highly sophisticated strategies to solve it must be developed and applied. Thus, most of the graph theoretical algorithms are highly adapted to a very special set of constraints. Often they are optimized, to cope with exactly these boundary conditions as far as possible.

The derivation of an efficient frequency plan for a particular broadcasting service is a dynamic process. This means that constraints are very likely to change in the course of the generation of a plan. Hence, optimal adaption of an algorithm to a specified set of constraints clearly is a major obstacle when deriving a good frequency plan. A method that works very well for a certain type and number of constraints need not necessarily be as efficient if the basis changes.

Stochastic optimization algorithms have undoubtedly advantages in that respect. In contrast to dedicated graph coloring algorithms, they constitute a general framework that is quite often copied from strategies developed by nature when trying to solve a problem or adapt to a particular environment. Evolution is a prominent example for a successful strategy to cope with very complex optimization problems. All these approaches can be cast into a set of simple rules that are iteratively applied. In most cases, these simple rules can be easily cast into an simple algorithm that allows to easily adapt to different optimization scenarios. Clearly, the price to pay is that a flexible approach can never be as efficient as a specially designed method.

In practice, one of the major problems frequency planning must face is the scarcity of available frequencies. As a consequence, transmitters must share a channel, even though this leads to high mutual interference levels. A natural objective when setting up a frequency plan, thus, is to reduce interference as much as possible. To this end, an objective function needs to be defined, which allows to assess the quality of a frequency plan in quantitative terms. One possibility thereto is to combine all mutual perturbations to be expected between different allotment areas in a analytical way in order to generate a global quality criterion. Exploiting this criterion allows to distinguish between good and bad frequency plans.

5.8 Frequency Planning Scenarios

In Section 5.6, the general outline of the generation of a frequency plan has been given. However, the term "frequency plan generation" should not only be understood as the process of assembling a complex frequency plan by means of an international conference. This is certainly covered by that phrase but it refers to the most important activity in the field of frequency planning only. Before such an enterprise can be accomplished, more limited tasks need to be managed. They can be divided into two distinct cases. On one hand, this concerns the question how much spectrum is needed for a certain service under given conditions. On the other hand, in reality frequency planners are usually confronted with the fact that there is not enough spectrum available in order to satisfy all requirements.

5.8.1 Spectrum Demand Studies

The successful introduction of radio or telecommunication services has to meet several technical and nontechnical constraints. The technical side deals with the question which transmission technique is utilized to provide which service. Nontechnical issues are economic feasibility and efficiency. This is directly linked to the question about the coverage target. What special services are to be provided in which areas and how much of these can be implemented in accordance with consumer demands and their will to spend some money for these new services?

In the case of radio and television, economical considerations are only one side of the coin. Usually, the introduction of broadcasting services has a political dimension, too. The reason is due to the role public broadcasters play in the field. They are liable to national or regional governments and their corresponding administrations under rules defining precisely the task and scope of public broadcasting.

Both political and economic demands thus influence the introduction of radio and telecommunication services. Once a road map for the introduction and the future development has been settled, the question arises how much spectrum is necessary to bring the considered service into operation. This calls for an appropriate estimation of the spectrum demand.

Basically, the fundamental assumption for a spectrum demand study is that there is an infinite amount of spectrum available. A set of requirements given in terms of allotments or assignments is subjected to a compatibility analysis as described in Section 5.6.1 in a first step. Then, based on the results given in terms of an adjacency matrix, frequencies or channels are assigned to the requirements that are virtually taken from an infinite pool of available channels. However, the frequency assignment is so carried out such as to minimize the number of different channels employed. The minimum number of channel then constitutes the spectrum demand for the set of requirements under the conditions imposed during the compatibility analysis. Algorithms as described in

Section 5.7 or others can be applied to calculate the actual frequency assignment.

In 2000 WorldDAB published a spectrum demand study for T-DAB on the occasion of the 9th CEPT conference in Lisbon [Wor00] (see also [Beu01]). The objective of the study was to learn how much spectrum might be needed in order to map the existing FM radio landscape into T-DAB. By means of concentrating on several representative regions in Europe a method had been proposed, which allowed realistic spectrum demand analyses. In detail, the environments of Paris, London, Rome, Stuttgart, and Malmoe were considered. The dimensions of the geographic regions to be investigated were different in all cases, whereas the German region was the largest.

The selection of examples should cover as much as possible different aspects that are important for the spectrum demand investigation. In France at that time only L-Band T-DAB blocks could be used, while in Great Britain and Sweden only the VHF range could be employed for T-DAB. In addition, the southern part of Sweden is surrounded by sea so that effects due to wave propagation above water had to be taken care of.

The purpose of the study was not only to estimate the amount of spectrum that would be required for a complete mapping of the actual FM radio landscape to T-DAB. Indeed, a forecast with respect of the future spectrum demand should be attempted, too. Clearly, this implied to make comprehensible assumptions with regard to the potential progress of the T-DAB system during the next ten years.

It proved to be very difficult to find any reliable information thereto on which a reasonable forecast could be based. There were still too many open questions concerning the future market penetration of T-DAB. Therefore, all estimates presented had to be considered as first vague indications only. However, the general assumption was that T-DAB is going to be a success thereby leading to an increasing demand for transmission capacity on the part of the broadcasters.

A set of polygons has been generated each of them representing the service area of one single FM radio program. In the first step, it has been assumed that six FM programmes should be combined to build one T-DAB multiplex. This decision is obvious because the total data capacity of a T-DAB multiplex allow the transmission of six audio signals of 192 kBits/s. This corresponds to stereo programmes in near CD quality. Clearly, the number of programmes within a multiplex could be varied, too. A figure between four and eight programmes seems to be reasonable. It has to be emphasized, however, that accepting this additional degree of freedom will further increase the complexity of the problem.

The investigation started by bundling six FM programmes to represent T-DAB multiplexes, respectively. The associated services area of the multiplex has been assumed to be given by the union of the individual programme service areas. That way a single allotment had been generated. Subsequently, a graph coloring strategy has been applied to determine the corresponding spectrum demand for that particular set of allotments. A full description of the T-DAB spectrum demand study can be found in [Wor00] or [Beu04b].

The definition of spectrum demand study given above is straightforward. In practice, a slight modification thereof might be more interesting. Frequency planning conferences, which are convened to set up corresponding frequency plans do not deal with an infinite amount of spectrum. Usually, a certain frequency range is defined out of which frequencies can be selected for the frequency assignment process. What is important during the preparation phase of such conferences is to get a clear idea how much requirements can be accommodated in the spectrum range at hand under idealistic conditions. This gives an overview about the potential usage of the spectrum.

A quite natural way to investigate this is to work on the basis of national layers of coverages. Even though there is no official definition of a coverage layer, there is an intuitive understanding of this concept. In the context of allotment planning, a national layer can be understood as the totality of allotments whose areas cover the national territory of a country completely. Figure 5.20 shows a simple example in the case of Germany.

The allotment areas shown in Figure 5.20 correspond to allotment areas that have been used for DVB-T in Germany during the preparation of the RRC-06. Before the conference it was important for German broadcasters to get an idea how much DVB-T allotments could be accommodated in the available spectrum, for example, in the UHF range. To this end, simulations have been carried out based on sets of allotments like in Figure 5.20 to calculate the spectrum demand. Several layers have been constructed by making a corresponding number of copies of the basic set of allotments. The total set of allotments has then been fed into a graph coloring algorithm to calculate the number of required UHF channels. It turned out that 6–8 channels are needed to satisfy

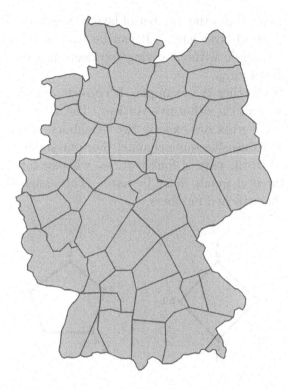

Figure 5.20: Allotment areas forming one national layer for Germany.

a single layer having a structure like that in Figure 5.20. Clearly, this result depends on the typical size and shape of the allotment areas under consideration as well as on the reuse distance for co-channel usage. Since the UHF range comprises a total of 49 channels between channel 21 and 69 roughly seven layers for DVB-T can be accommodated in UHF. This assessment has been confirmed by the results of conference [ITU06].

The spectrum demand of several identical layers of allotments can be easily estimated by calculating the demand for one layer and then multiplying this result with the number of layers. However, it should not be forgotten that the underlying mathematical problem associated to such a simple spectrum demand estimation is very complex. Therefore, it should not come as a surprise that solutions to that mathematical problem exist, which show interesting characteristics. Indeed, it is possible to find solutions that require less spectrum than the simple formula "demand for one layer times the number of layers" suggests. This is usually referred to as "multi-layer gain." It can be explained with the help of a very simple example without diving too deep into the mathematical basics of graph coloring.

Figure 5.21 sketches the example. It consists of two simple graphs, one having four and the other five vertices. The vertices are labeled by characters. Each vertex has exactly two neighbors, which consequently are not allowed to use the same channel. For one single layer two channels are needed in the case of four vertices, whereas for the graph with five vertices three channels must be used. The assigned channels are indicated by the attached numbers.

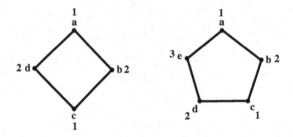

Figure 5.21: Two simple graphs representing a single coverage layer consisting of four and five allotment areas, respectively. The characters denote the vertex while the numbers represent the assigned channels.

Figure 5.22 shows a solution for the corresponding two layer scenario. Speaking in terms of allotment areas this would correspond to a duplication of the already existing allotments. Thus, a valid solution for this graph coloring problem is a duplication of the number of required channels. In the case of four vertices there is no other possibility than that. However, the structure of the graph having five vertices allows for a solution that needs five instead of six channels as obtained from the simple duplication scheme.

Compared to real world problems in the field of frequency assignment and spectrum demand estimation these examples indeed have to be considered as being trivial. However, if already under such simple and transparent conditions multilayer gain effects can be observed it seems

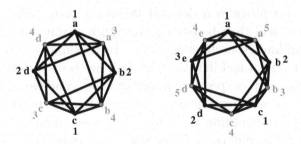

Figure 5.22: Two simple graphs representing two coverage layers. The characters denote the vertex while the numbers represent the assigned channels. The two layers are differentiated by black and gray color.

to be quite obvious that in the case of more complex problems such a degree of freedom can be expected just waiting to be exploited.

5.8.2 Constrained Frequency Assignment Problems

The estimation of the spectrum demand is a very important issue with respect to the introduction of a new broadcasting service. Knowledge of the spectrum demand is a prerequisite in order to decide if the envisaged planning targets are realistic or need to be revised. The scenario discussed with respect to Figure 5.20 above, namely the question how many DVB-T layers can realistically be accommodated in the UHF spectrum is good example thereto.

Nevertheless, such an investigation is just the prelude for the task usually dealt with in terms of international conferences, which is the generation of a frequency plan for selected services. In that case, however, the question to be answered is under what conditions is it possible to accommodate a given set of requirements in a limited amount of spectrum. This leads to frequency plans where not all requirements can be satisfied. This is equivalent to asking how much interference has to be accepted if a certain subset of the entire input requirements is allocated to the same channel, and are the levels of interference acceptable to allow for a proper operation of the corresponding broadcasting networks?

The result of the spectrum demand estimation discussed in the previous section not only gives an answer about how much channels are needed but also provides a detailed frequency plan. This means that each allotment will then be associated with a channel. Clearly, such a plan is only of limited usefulness because two assumptions have been included which are not met in reality. First, it has been assumed that there is an infinite amount of spectrum available and second Germany has been treated as being independent from any neighboring country. In particular, the latter assumption is wrong. Since wave propagation does not stop at national boundaries neighboring countries are never independent as long as one of them does not make use of the considered frequency range at all. Therefore, the results of the spectrum demand estimations always have to be considered as very rough guesses for the number of channels that is needed rather than providing reliable frequency plans.

Constrained frequency assignment problems come in very different variations. The task the RRC-06 had to carry out leading to the GE06 Agreement [ITU06] is a prominent example. This very important planning conference is discussed in more detail in Chapter 8. For the time being it suffices to say that the scope of the RRC-06 was to establish a frequency plan for DVB-T and T-DAB for a very large planning area taking into account an awful amount of very different constraints. This refers to the availability of spectrum that differed across the planning area of the RRC-06 dramatically, the number of requirements administration wanted to include in the frequency plan, and also the existence of other primary services like radio navigation, which had to be protected during the plan generation process.

In connection to the preparation of the RRC-06, groups of administrations developed coordinated input requirements to the RRC-06. To this end, frequency planning exercises have been conducted on a multicountry level. This means that a frequency plan has been calculated, for example, for Germany, Netherlands, Belgium, Luxemburg, France, and Switzerland under the assumption that in the countries adjacent to these countries sets of allotments have been incorporated having fixed channel allocations for all allotments. This way an external boundary condition has been constructed, which then could be explicitly taken into account when assigning channels to allotments in the countries lists before. Clearly, these frequency assignments were carried out such as to minimize the number of conflicting co-channel allocations.

In order to carry out such investigations, a special software tool has been developed based on stochastic optimization principles [Beu04a].

Actually, the so-called "Great Deluge Algorithm" is employed [Due93]. It is described in more details in Annex A.1. This algorithm proved to be very robust and reliable. Furthermore, it is very easy to incorporate different constraints and offers therefore a lot of freedom in order to set up planning exercises. It is evident that other optimization strategies based on other stochastic optimization algorithms can be adopted as well. However, it is important to understand that the type of problems discussed here can be solved in terms of stochastic optimization tools more efficiently than by graph coloring methods.

The starting point of every constrained frequency assignment based on allotments is a set of allotment areas. In principle, they can have arbitrary shapes. They are approximated by correspondingly chosen polygons. In order to keep the administrative effort as small as possible, the number of vertices that can be used to describe an allotment area is usually limited. The specification of the number and the location of the vertices defines the geographical shape of an allotment area. Usually, geographical coordinates are used to represent the vertices. They can be easily converted into other coordinate systems that might be better adapted to cope with regional or even local conditions. For practical reasons, allotment areas are supposed to posses a simple geometrical structure. In mathematical terms, this means they should preferably be simply connected geometrical objects. Holes or loops are to be avoided. Allotment areas are usually allowed to partially or totally overlap.

Constrained frequency assignment means in the first place that access to spectrum is restricted. This can refer to a global restriction in the sense that within a country certain frequencies are not available for a particular service at all. But it can also mean that the access to spectrum differs from allotment area to allotment area. A typical example is a situation along the border of a country. An administration has no direct influence on the spectrum usage in the neighboring country, and therefore certain channels might not be accessible in certain areas or regions along the common border. Consequently, the definition of an allotment area has to be extended by specifying a list of all frequencies that can be used within that area. During the frequency assignment process, the allotment is assigned only one of those frequencies that are contained in its list. That way it is assured that any assignment is valid at least as far as the usage of the frequency resources is concerned. Any frequency plan derived, therefore, does not include any cases where a frequency cannot be used.

As in the case of spectrum demand studies, also for constrained frequency assignment the solution is obtained by a two-step process consisting of compatibility analysis and frequency plan synthesis or generation. The compatibility analysis is identical to that of the spectrum demand case. If a reuse distance based approach is applied, the effective distances between allotments have to be calculated for all possible pairs. The result of the compatibility analysis is given in terms of a adjacency matrix introduced in Section 5.6.1.2 of which entries say if two allotments can share a channel or not. Typically the value "0" indicates sharing is possible, while "1" marks an incompatibility.

The value of the reuse distance that is used in order to evaluate the compatibility between two allotments is calculated on the basis of several assumptions. In Section 5.5 the standard way is explained. However, it might turn out that the reuse distance value inaccurately represents the interference relation between two allotments. It has to be borne in mind that the concept of allotments combined with RPCs and RNs is a theoretical construct. Real world network implementations might exhibit significantly different characteristics and network planners might be well aware that the transmitters foreseen to be brought into operation in the two allotments at hand will experience no channel sharing problem even though the calculated effective distance is less than the reuse distance. The reversed situation might be encountered as well in practice. Two allotments might be separated by more than the reuse distance but the network implementations will be based on transmitter sites that cannot share frequencies due to special topographic conditions. Such situations can be accounted for by introducing compatibility or incompatibility flags between allotments. Whenever such a flag is found during the compatibility analysis, its result for a given pair of allotments is discarded and the two areas are either considered as compatible or incompatible.

As mentioned above, constrained frequency problems are preferably tackled in terms of stochastic optimization algorithms rather than by graph coloring methods. Stochastic optimization requires to define an appropriate objective function. This objective function must depend on the actual allocation of frequencies to the set of considered allotments in a defined way.

In general terms, stochastic optimization is an iterative process. This means a very large number of frequency plans are calculated, and according to given rules, the best plan is chosen. Calculation of frequency plan means to assign to each of the allotments under consideration one of its permissible frequencies. Such an arbitrary assignment will in most cases lead to a very inefficient frequency plan full of incompatibilities, and thus a high level of interference.

Any other arbitrarily chosen frequency plan might differ in detail but the probability that it suffers from the same problems is very high. Therefore, two questions arise. First, how can the quality of a frequency plan be assessed, that is how can "good" plans be distinguished from "bad" plans? The second question coming up immediately is how can a good frequency plan be derived or calculated?

The latter question can be answered more easily. The derivation of a good frequency plan corresponds to a very complex combinatorial problem from a mathematical point of view. Due to the large number of individual constraints to be taken into account stochastic optimization algorithms can be applied to find an optimal solution.

The former question concerning the quality needs further specification. First of all, it must be possible to define an attribute like "good" in the context of a frequency plan at all. This means a precisely defined quality function must exist. A first requirement thereto is the uniqueness of this function. A particular frequency plan can only have one single quality.

Moreover, the mathematical form of the quality function depends on several factors. Clearly, the number of incompatibilities will certainly be important. This alone, however, will not be sufficient. If two allotments can share a channel or not is a black-or-white statement, this does not give enough freedom to fine-tune the objective function. More information about the compatibility relation between two allotments is needed. Basically, the effective distance that is calculated during the compatibility analysis provides such information. Its value can be employed to define the contribution of each pair of allotments to the objective function. If two allotments have been given the same channel and at the same time are separated by less than the co-channel reuse distance, then it is obvious that there is difference, if the effective distance between them is 119 km instead of the required 120 km or only 80 km instead of 120 km. So, the explicit value of the effective distance between two co-channel allotments can act as direct measure for the interference level to be expected.

Furthermore, there will always be overlapping allotment areas, either partially or totally. This is a direct consequence of the fact that at a given geographical point more than a single broadcasting service will be offered. If however, different network structures are utilized for the implementation of overlapping allotments, adjacent channel interference problems are likely to occur. In principle, this can always be resolved by using the same transmitter sites for overlapping allotments. Sometimes, however, this is not feasible because the networks are operated by different network operators using their own transmitter networks. In case, this is unavoidable or even part of a national strategy of an administration, then this issue should be taken into consideration during the frequency plan generation already.

Similar to the way in which the effective distance between allotments can be exploited to define a measure for the expected interference between allotments, the spectral distance between the corresponding frequency allocations has to be considered appropriately. The first step would be to define a critical spectral distance Δf_c beyond which no problems will occur in overlapping areas. Figure 5.23 illustrates a simple scheme to identify adjacent channel or co-channel situations.

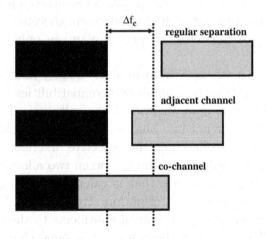

Figure 5.23: Simple scheme to define adjacent and co-channel usage of spectrum. The black and gray rectangles are to represent the spectrum ranges used by two networks in overlapping areas.

Clearly, if the spectral distance is to be taken into account during the assessment of a frequency plan, the compatibility analysis has to provide the information about potential overlap between allotment areas. It could be even relevant to distinguish between partial and total overlap, for example, by specifying the amount of overlap. This could then be incorporated into the objective function appropriately.

Effective distance of co-channel allotments and overlap of adjacent channel allotments determine the contribution of a pair of allotments to the value of the objective function. Qualitatively the negative impact they might have can be visualized as shown in Figure 5.24.

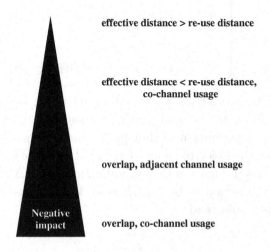

Figure 5.24: Impact of the adjacency relations between two allotments on the objective function.

A first and obvious approach to quantify the incompatibilities between two allotment areas i and j is to define their incompatibility Δ_{ij} as the difference between the corresponding reuse distance and the effective distance of the two areas. In mathematical terms this can written as

$$\Delta_{ij} = RU - ED(i, j), \qquad (5.8)$$

where RU denotes the reuse distance above land and $ED(i, j)$ the effective distance between areas i and j.

In the case of overlapping areas, the effective distance clearly is equal to zero. However, it might be necessary to distinguish incompatibilities according to the degree of overlap. A quite natural measure is to calculate the fraction ψ of the overlap with respect to the total area of the two coverage areas, for example,

$$\Psi_{ij} = \frac{2 \times \text{size of overlapping area}}{\text{area}(i) + \text{area}(j)}. \tag{5.9}$$

Then, the definition of the incompatibility Δ_{ij} could be extended, for example, in the form

$$\Delta_{ij} = \begin{cases} RU - ED(i,j) & , & ED(i,j) > 0 \\ \\ RU + 100 \times \psi_{ij} & , & ED(i,j) \leq 0 \end{cases}. \tag{5.10}$$

This way the overlap of allotment areas using adjacent channels would be accounted for in quantitative terms. The larger the overlap is the larger the contribution of that particular pair of allotments is.

A straightforward idea to generate a global measure for the quality of a frequency plan starting from the individual measures Δ_{ij} is to sum them all. The quality function to be exploited by the stochastic optimization algorithms thus would read

$$\Delta = \sum_{i,j} \Delta_{ij}. \tag{5.11}$$

The definition (5.8–5.11) has to be considered only as one special way to handle this. In practice, it might turn out that a different formulation has to be found, because the difference of the quality of two frequency plans does not allow to objectively decide which one is better. Quite often it is necessary to declare the overlap cases of Figure 5.24 as forbidden to suppress the occurrence of such incompatibilities with high probability. The quality "forbidden" can be realized in terms of a predefined very large value for Δ_{ij}, so that the penalty for accepting the corresponding assignment in the plan will be extremely high. Figure 5.25 sketches a decision tree in order to assess the incompatibility between two areas for such an approach.

The decision process starts with the calculation of both the effective distance and the spectral distance between the two areas under consideration. Then the incompatibility Δ_{ij} is calculated along the lines of the decision tree of Figure 5.25. According to Equation (5.11) all individual contributions are summed to obtain a global measure for the quality of

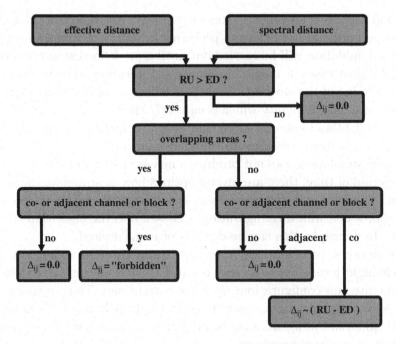

Figure 5.25: Decision tree to pass through in order to assess the
incompatibility between two allotment areas.

the current frequency plan. If necessary, appropriately chosen weights
can be used for the summation to put more emphasis on particular sections of the frequency plan.

The generation of a frequency plan by means of stochastic optimization usually starts with an arbitrary initial frequency plan. This is accomplished by choosing for each of the allotments one if its allowed frequencies by chance. Subsequently, this initial plan is assessed by passing
through the decision tree of Figure 5.25 for each pair of allotments. Then
the optimization cycle can be started.

During the simulation a new configuration is generated from the current one by arbitrarily choosing one allotment and assigning another
of its allowed frequencies also by chance. Clearly, it is also possible to
change more than one allotment during one iteration step. It is reasonable to try different strategies in order to obtain the best results. Such
freedom is typical for stochastic optimization algorithms. There are no
precise rules how to proceed in a given situation. On that level, the

solution process is governed to some extend by trial and error. However, this does normally not pose any problems in practical applications.

The simulation will be carried out until a predefined stop criterion is met. In most cases, a maximum number of iteration steps is given. At the beginning of the stochastic search usually a lot of new configurations are found having a quality which is currently the best. As the simulation proceeds it takes more and more time to find a new best solution. That way, the algorithm converges to the final solution.

If the simulation does not produce a new best solution after a reasonable period of time, there are several options how to stimulate the search again. The most straightforward way is to restart the whole process. To this end, a new initial configuration is chosen and the algorithm started again. In principle, this can be done as often as desired.

Sometimes, however, it is better not to start from scratch again. The search might have come to an end, because there is not much freedom left to generate new configurations from the current one. Thus, relaxing this a little bit and allowing for more freedom might help to escape from the dead-end situation again. Like before, such a relaxation can be applied as often as it seems necessary.

During the preparation of the RRC-06 many simulations concerning different planning scenarios have been carried out. There have been a enormous number of bi- and multilateral meetings of administrations and broadcasters in order to align their requirements for spectrum. Studies based on the stochastic optimization have been one of the pillars to find solutions all concerned parties could agree to in the end. More details and examples can be found in [Beu04b].

Chapter 6

Network Planning Basics

Frequency assignment and network planning can be considered as two sides of the same coin. They are independent from each other on one hand but at the same, they are closely linked. The generation of a frequency plan rests on the technical characteristics of the networks that are to be implemented later, that is how much interference a typical network is likely to impose on other networks and vice versa, how a network can be implemented in the presence of interference from outside. These characteristics determine parameters like the co-channel reuse distance, which in turn has a crucial impact on the spectrum demand.

On the other hand, before a network can be brought into operation a frequency resource must be provided. In addition, the conditions under which this can be accomplished have to be defined in detail. For example, two neighboring administrations might have agreed that the implementation of a particular entry in a frequency plan is subject to other criteria—more relaxed or also more stringent ones—than the formal application of coordination procedures associated with a corresponding frequency plan would request.

6.1 Planning Parameters

Planning a terrestrial broadcasting network usually refers to the task of providing a certain broadcasting service throughout a large area. There are frequency ranges where a single transmitter is able to provide service for an entire country and beyond. However, if VHF or UHF frequencies are employed, the wave propagation conditions prohibit such enormous coverage areas. As a consequence, several transmitters need to be utilized

R. Beutler, *Digital Terrestrial Broadcasting Networks*,
DOI 10.1007/978-0-387-09635-3_6, © Springer Science+Business Media, LLC 2008

to cover a whole country. Depending on the technical characteristics of the terrestrial broadcasting system, a set of different frequencies has to be used in order to avoid unacceptable interference between the transmitters. Hence, planning a broadcasting network is tantamount to dealing with several transmitters at the same time.

In order to successfully plan a terrestrial broadcasting network an appropriate model of the whole network has to be defined. Each model of a transmitter network rests on four pillars. First of all, the set of transmitters forming the basis of the network have to be specified by fixing their relevant technical parameters. Then, an adequate wave propagation model is needed that allows to calculate the field strength a transmitter produces at a given point of reception. Combination of different signal contributions is an issue that has to be considered in that context as well. Furthermore, the characteristics of the receivers under multipath propagation conditions must be known at least qualitatively. And last but not the least, the target service area is to be defined appropriately.

In the analogue world, FM radio and in particular analogue TV have been planned for stationary reception using a directional antenna. The customer was expected to accept some technical and financial expenditure to have a good reception quality. Therefore, all planning activities assumed a receiving antenna height of 10 m above ground and an antenna gain of 6 dB.

At the time FM radio was introduced, it could not be foreseen that mobile reception in vehicles will become as important and common as the standard roof antenna reception. This is also true for portable indoor reception. Since the receiver manufactures made large progress in terms of providing steadily increased receiver quality both mobile and portable indoor reception are considered as being satisfying today.

However, this is not true for terrestrial analogue television. Still a directional antenna is the prerequisite for acceptable reception quality. It is true that in the vicinity of a television transmitter, it is possible to use small portable television receivers equipped with a short rod antenna. However, mobile reception in vehicles usually does not work.

For digital terrestrial broadcasting systems several reception conditions have been considered from the very start of the introduction of these systems. Still there is a need for fixed-roof top reception with an directional antenna. But at the same time other reception modes like portable outdoor, portable indoor, mobile, and handled reception are very important, depending on the system employed and the characteristics of the content delivered. For T-DAB and DVB-T the corresponding planning parameters like minimum field strengths and protection ratios for all possible interactions between system variants have been investigated in detail. The details can be found, for example, in [CEP95], [CEP96], [CEP97] and [ITU02].

Generally, FM radio and analogue TV have to be planned as multi frequency networks. This means that in order to offer one particular programme throughout a large area like a whole national territory a multitude of different frequencies has to be utilized. In contrast, digital terrestrial broadcasting systems like T-DAB or DVB-T can be operated as single frequency networks (SFN) due to the technical characteristics of the employed COFDM technology. In that case, a set of transmitters broadcasts the same content, using the same frequency or channel. The details of the network planning process certainly differ between analogue and digital systems. But, in both cases several transmitters have to be dealt with at the same time in order to provide services throughout an extended area. In the following sections, the most important elements of network planning for digital terrestrial broadcasting systems are introduced. An extensive description of SFN planning with a slightly different focus can be found in [EBU05].

6.1.1 Network Parameters

In order to plan a broadcasting network, the characteristics of the transmitters have to be defined in detail. A transmitter network is built from a set of N transmitters. They are supposed to be located at the positions $\mathbf{r}_k$ ($k = 1, .., N$). Each of the transmitters is operated at an individually adjustable radiation power. The antenna is fed with the input power P_I, which is then transformed into the radiation power P_T according to the efficiency $\eta = P_T/P_I$. The quantity η accounts also for cable losses. The transmission generates a power density $S(\mathbf{r})$ at a distant point $\mathbf{r}$, whose value depends on the distance to the antenna and on the topographic and morphologic conditions.

Another important factor determines the field strength in practice, too. Every antenna employed gives rise to a characteristic three-dimensional angular antenna pattern. The most simple antenna pattern is that of a three-dimensional isotropic radiator. In that case, the radiated power is released uniformly in each direction.

However, spherical symmetry is not desirable for broadcasting purposes. The transmission should preferably take place in horizontal direction. A slight tilt towards the earth's surface is even better. Furthermore, most transmissions are subject to certain restricting constraints in the sense that the output power towards specified directions must not exceed limits imposed by the presence of other radio services.

With the help of proper antenna design, it is possible to bundle the available power towards defined directions at the expense of others (see for example [Bal82] for a detailed discussion). This non-isotropic transmission is described in terms of the so-called antenna gain. It is defined as the ratio between the power density S of the considered antenna into the boresight direction and the power density of an ideal isotropic radiator S_i, under the condition that both are virtually fed by the same input power P_I.

Radio or television antennas are usually based on dipoles, a set or dipoles, or related designs. Therefore, it is reasonable to relate the antenna gain not to the isotropic radiator but to the power density S_d of the main lobe of an ideal half wave dipole [Kat89]. The definition of the antenna gain used in broadcasting, thus, reads

$$G_d = \frac{S}{S_d}. \tag{6.1}$$

The antenna gain is a global characterization of an antenna. Basically, it is a first indication for the directivity of the antenna. To provide the full information about the angular behavior it is, however, necessary to make reference to the complete antenna pattern or diagram. Usually, two types of antenna patterns are given, one in the horizontal plane and one in the plane perpendicular, thereto. The values of the antenna diagrams represent a measure for the power emitted into a particular direction. They are normalized to the boresight direction. Thus, in logarithmic representation only negative values appear.

To establish a unique characterization of the transmission power of different transmitters, the half wave dipole is utilized thereto as reference as in the case of the antenna gain. The so-called effective radiated power (ERP) is just the product between the input power and the antenna gain:

$$\text{ERP} = P_I * G_d \ . \tag{6.2}$$

Usually, logarithmic representation is used for the ERP as well,

$$\text{ERP}_{\log} = 10 * \log(\frac{\text{ERP}}{P_{\text{ref}}}),$$

where for the reference power P_{ref} both 1 kW as well as 1 W are used in practice. The statement of 3 dBkW, thus corresponds to an ERP of 2 kW.

The isotropic radiator mentioned above cannot be realized as transmitting antenna. Even an antenna, whose pattern shows a circular geometry in the horizontal plane cannot be put into practice without tremendous effort. Significant deviations from the circular shape have to be accepted usually. Also the implementation of an exact half wave dipole remains an exception.

Real antennas consist of dipoles and Yagi antennas that are combined appropriately to approximate the desired theoretical antenna pattern as good as possible. Exact antenna diagrams are usually not known in practice. As a rule, numerical simulations are carried out for a particular antenna configuration to obtain realistic values for the antenna patterns.

Sometimes, if the effort is worthwhile, for example, in the case of antennas of high-power stations, measurements with the help of helicopters are carried out in order to improve the knowledge about the actual transmission characteristics. Experience showed that unfortunately in most cases there is a large discrepancy between antenna patterns employed for theoretical planning and those actually employed in practice. Therefore, it has to be borne in mind that theoretical antenna diagrams have to be considered as rough approximations to the real patterns only.

Usually, for network planning purposes only horizontal antenna diagrams are taken into account. Vertical patterns are considered only for special case studies. The angular dependence of the horizontal radiation patterns are approximated by specifying 36 parameters A_k attached to angles from $0°$ to $350°$ separated by an interval of $10°$. The north direction coincides with the angle $0°$ increasing with mathematically negative orientation. For angles that are not equal to a multitude of $10°$ a linear interpolation is applied. Figure 6.1 shows a typical diagram that is used for planning.

There are two basic constraints to be taken into account when generating theoretical antenna diagrams. In practice, reductions beyond 20 dB cannot be realized without very large effort. When bringing into operation a transmitter its compatibility with already existing transmitters

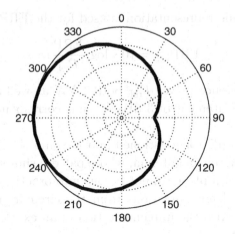

Figure 6.1: Example for an antenna diagram used for planning.
The scale is 2 dB, so the maximum reduction is about
6 dB towards east.

has to be checked. If during the corresponding coordination procedures
reductions of that order of magnitude result, the only solution is to reduce the total ERP of the considered transmitter in order to obtain a
corresponding reduction.

The second constraint to be met when designing theoretical diagrams
is to take care that the reduction as a function of the angle does not
exhibit too strong changes, that is the corresponding gradient needs to be
sufficiently small. Cast into practical terms, this means that a difference
between adjacent antenna diagram values must not exceed 5 dB.

In contrast to analogue transmitter networks there exists another
important parameter in the case of T-DAB and DVB-T single frequency
networks that primarily determines the reception quality. This crucial
feature is connected to the way COFDM systems can cope with the
special characteristics of multipath transmission channels. Typically,
under such conditions more than one signal contributions arrive at the
point of reception (see Section 4.2).

In the case of FM radio, two signals arriving at different times at
the receiver will lead to service degradations due to their interference.
This cannot be overcome by the FM receiver. Only in the very unlikely
situation do they arrive exactly at the same time and they would add
constructively. Just a very small deviation from the contemporaneous

arrival creates perturbations. The larger the difference between their times of arrival becomes the more serious the problems gets.

For digital terrestrial broadcasting systems based on COFDM, this problem does not exist up to certain limits beyond which problems appear. However, the systems design was carried out exactly so as to compensate for typically occurring time of arrival delay spreads. The introduction of a guard interval allows the "constructive collection" of different signal contributions.

If the time of arrival of different signals at the location of the receiver is no longer a critical issue, then conversely it is no longer necessary to exactly synchronize all transmitters participating in a SFN in time. The actual transmission instant of the transmitters is allowed to vary within certain limits. But this gives rise to the possibility to exploit this degree of freedom by configuring the network correspondingly. Thereby, it is possible to eliminate the occurrence of self-interference problems.

Such situations might be encountered, if there is a bounded part of the total service area in which due to special topographic conditions a distant transmitter still can be received. If its time of arrival is such that the difference with respect to the times of the others is larger than the guard interval, self-interference would result. However, by advancing the transmission instant of that particular transmitter a delay spread could be configured that would be harmless. Rather, another constructive signal contribution would result.

In principle, such a type of configuration could be applied to any transmitter making part of the SFN. Consequently, in order to characterize a transmitter completely it is necessary not only to describe it in terms of power and antenna diagram but also to include an individual transmission delay τ_k, too.

Apart from pure technical parameters, realistic network planning necessarily must deal with economic aspects as well. The erection and the operation of a transmitter is a very expensive enterprise. Several factors have to be considered. First of all, an adequate property to build a tower is needed. Maintenance facilities are to be provided as well. Furthermore, an expensive technical infrastructure must be set up. This covers antennas, cable connections, amplifier, transmitter unit, and much more.

Expenses related to local infrastructure is one thing, another issue is the way in which the signal that is to be broadcast is delivered to the transmitter location. According to individual local conditions data communication lines, satellite links, or fixed radio-link systems can be

utilized thereto. Each of them requires different technical equipment and
can be employed under different conditions only.

All these factors have to be included into a proper calculation of
the costs for setting up and maintaining a transmitter site. Therefore,
quite naturally in practice, it happens that although a particular coverage target has been defined a priori, the calculation of the costs forces
broadcasters to decide that this objective cannot be realized. Applied to
the problem of network planning, this means that it would be reasonable
to include the costs K_k a transmitter causes into the planning process
from the very beginning.

Certainly there are other aspects that could be considered in order
to model the characteristics of the transmitters. However, the principal
ideas about network planning are not affected by reducing the network
parameters to the set of quantities discussed so far. Table 6.1 summarizes
those parameters that are used for the planning examples here.

To conclude this section, one point not mentioned so far needs to
be emphasized. In addition to the set of parameters given in Table 6.1,
usually the description of a transmitter site is supplemented by the specification of 36 effective height values (see Section 4.3.1). They are to be
interpreted the same way as it is done for the antenna factors, that is
they represent the values for the directions 0°, 10°, 20°, etc. As with the
quantities A_k representing the antenna diagram, a linear interpolation is
applied for the effective heights at angles in between as well.

Table 6.1: A set of network parameters employed for planning
purposes.

$\mathbf{r}_k$:	Location vector of the k-th transmitter site
P_k :	ERP of the k-th transmitter
A_{km} :	Antenna factor of the k-th transmitter towards direction m
τ_k :	Time delay of the k-th transmitter
K_k :	Costs of the k-th transmitter

It is evident that this information is essential for the application of the wave propagation model according to the ITU recommendation 1546 [ITU01a]. However, this parameters are fixed as long as the location of the transmitter is fixed. If the position of a transmitter is changed, these parameters have to be calculated again. Their values are linked to the topographic conditions and thus cannot be adjusted freely. Therefore, they cannot be considered as free parameters in the literal sense and hence they are neglected here.

6.1.2 Wave Propagation

Network planning for a terrestrial radio or television system rests on coverage predictions as described in Section 4. To this end, an adequate wave propagation model is needed that allows the calculation of the electromagnetic field strength at a specified point of reception. Depending on the required accuracy, the coverage prediction might take into account digital terrain data and morphologic conditions or not.

In particular SFNs benefit from the fact that at a given point of reception not only one single signal is received but also a number of signals. These originate either from different transmitters or are caused by reflections. The superposition of all contributions can lead to better service quality. This fact imposes to take into account two important aspects with respect to the prediction of the field strength. One point is related to the occurrence of reflections or echoes and the other concerns the selection of reception points for which the field strength is predicted.

Reflections are a consequence of the electromagnetic wave propagation under typical topographical conditions where mountains or hills exist (see Chapter 4). If echoes are important for the received signal strength three-dimensional wave propagation models should preferably be employed.

The development of such models has been subject to scientific research for a long time already. In the meantime, there are several models available that differ in particular with respect to the run times of the corresponding software (see for example [Leb92] and [Gro95]). Due to the mostly unacceptable large computation times these algorithms cannot be used for wide-area predictions on a daily basis. Therefore, mainly two-dimensional wave propagation models are usually employed. They evaluate the topography only along the line connecting the location of the transmitter and the receiver (see [Oku68], [Cau82], [Mee83], [Gro86] or [Lon86]). Moreover, in relation to network planning it is necessary that the prediction of field strengths can be carried out for any arbitrary point of reception. Utilized wave propagation models need to allow that flexibility. This demand becomes essential in relation to the selection of the set of reception points that should represent the service area as discussed in Section 6.1.4 below.

As a matter of principle, it is not possible to predict the field strength for each mathematical point within a given area. Therefore, statistical methods are employed. Thereby it is possible to describe the variation of the field strength across an extended area on the basis of a set of distinct points of reception inside this area.

Current wave propagation models provide information about the field strength value produced by the network at a chosen point of reception. In view of the technical characteristics of COFDM transmission systems, it would be desirable if the output of the wave propagation model would comprise the details of the radio channel as a function of time of arrival and angle of incidence of the signals. This would certainly allow for a more accurate determination of the field strength to be expected (see Section 4.5.4).

However, relying only on the field strength can lead to completely wrong assessments of the reception quality as measurements have clearly shown. There are cases in which the predicted field strength is large enough to allow for good reception. Nevertheless, unacceptable quality degradations are experienced due to self-interference arising quite naturally in multipath environments.

On the other hand, the reverse situation may be encountered as well. The wave propagation model predicts a field strength that is considered as being not large enough for good reception. Nonetheless, there is good reception. Obviously, there are additional constructive contributions to the signal at the receiver site due to reflections. This is connected to the technical characteristics of digital broadcasting systems like very efficient error correction mechanisms. There is no simple link between the provided field strength and the reception quality. The magnitude of the received field strength certainly is a first indicator, since if it is too small then no error free demodulation of the received signals will be possible anyway.

Therefore, the assessment of the reception quality can no longer be based on the field strength alone as quality measure. A new appropriate quantity like the bit error rate before the application of the error correction algorithms [Beu98a] should be employed instead.

In order to incorporate the bit error rate, future prediction methods have to take into account all details of the transmission channel. In the case of single frequency networks, this means the number of impinging signals, their amplitudes, angle of incidence, and relative time of arrival need to be known. Clearly, this implies the demand for better wave propagation models with particular emphasis on three-dimensional models.

6.1.3   Receiver Models

In order to assess the service quality at a given point of reception, the properties of the receivers have to be taken into account explicitly. For analogue broadcasting systems, such as FM radio or analogue TV, the specification of the field strength is considered to be sufficient to allow an assessment of the reception quality. The received signal must be strong enough to be distinguished by the receiver against a noise background produced by several sources including the receiver itself.

For analogue systems, there is basically just one wanted signal contribution. Even a ground reflection arriving at the receiver location just an instant later leads to interference. This is well-know and can be experienced when receiving a FM signal in a car. When waiting in front of a traffic light the superimposition of direct signal and ground reflection can lead to severe quality degradation. However, moving the car about half a meter might allow the receiver to decode the FM signal perfectly again. Under more pronounced multipath conditions, the FM reception will be even more problematic.

The situation is different in the case of digital terrestrial broadcasting systems employing COFDM technology. The guard interval provides a measure that allows to take advantage of reflections up to a certain delay spread between the different signal contributions. However, this immediately gives rise to the questions how the receiver responds to a multipath environment and how the received signal is evaluated.

In order to demodulate and decode the transmitted information, the receiver has to synchronize to the received signal, in case of T-DAB or DVB-T to the temporal succession of individual symbols of duration T_S. The signal is evaluated by periodically recording a set of samples within a time interval of duration $T_W < T_S$.

If there is only one signal contribution, then the synchronization strategy of the receiver is obvious. The evaluation window has to be aligned with the arriving signal. However, in a multipath environment the way this synchronization process is carried out needs some care. In that case, there is a variety of different signal contributions arriving at the point of reception at different times.

If the time interval between the first and the last signal is smaller than the guard interval T_G, a system like T-DAB or DVB-T can benefit from the superposition of several signals. In case, the first and the latest arriving signals are separated by a time larger than the guard interval, both constructive and destructive effects will result.

Seen from a physical point of view, the set of different signals are linearly superimposed at the point of reception, according to their amplitudes and phases. Both quantities are subject to the details of the wave propagation from the transmitters to the receiver.

The correct way to deal with a multipath environment in terms of a wave propagation model would be to add all signals coherently, that is to fully take into account their amplitudes and phases. Then, the resulting signal would have to be evaluated. However, there is no wave propagation model so far that is able to provide the necessary information thereto.

First of all, the exact determination of the phases is not feasible. To make that statement clear, it is reminded as the rule-of-thumb that any object in the order of the wave length has to be considered as a potential scatterer for a propagating wave. In the VHF range wave lengths of 1–2 m are found. As a consequence, a topographic data base that is supposed to fully cover this would need to possess a resolution of less than 1 m. There are certainly data bases of that type available today, but it is completely impossible to rest wave propagation calculations for a distance of 100 km on such a data base. Computation times would definitely explode. Furthermore, the scattering process of the electromagnetic wave would need to be captured on all scales exactly. This is clearly not possible from a practical point of view, and therefore the principle inaccuracy of typical wave propagation models does not allow to calculate the phases of the waves.

Since the physically correct way to deal with a multipath environment is actually not possible, alternatives are required. The standard approach usually adopted operates on the level of signal powers and assumes the arriving signals can be virtually subdivided into two parts, a wanted or useful power C_k and a corresponding unwanted or interfering power I_k. The values of C_k and I_k depend on the time of arrival of the k-th signal relative to the position of the evaluation window. Figure 6.2 sketches the situation.

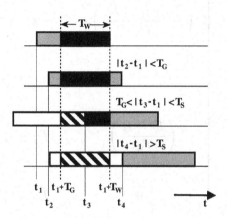

Figure 6.2: Constructive and destructive contribution of different signals depend on their relative time of arrival with respect to the position of the receiver evaluation window. Black areas contribute constructively, while the dashed parts give rise to selfinterference.

Looking at Figure 6.2 it becomes clear that all black parts can be counted constructively. In the case of the third signal, only a fraction can be added to the wanted signal. The rest increases the unwanted signal part. The different contributions to the wanted and unwanted signals are taken into account according to appropriate weighting factors. These weighting functions are different for T-DAB and DVB-T. The reason is that the processing of the signals by the corresponding receivers is slightly different. More details can be found in [EBU98] or [EBU03].

In order to apply the weighting functions, it is necessary to specify the synchronization strategy of the receiver. This means to specify how the receiver positions its evaluation window of duration T_W relative to the temporal arrangement of the incoming signals. In Figure 6.2 the receiver synchronizes to the first arriving signal. Without restriction the time t_1 can be put equal to zero. Then for T-DAB the weighting factors f_N and f_S for wanted and unwanted as a function of the relative time of arrival read

$$
f^{DAB}_{wanted} = \begin{cases} 0 & , \quad t_k < -T_W \\[2ex] \left(\dfrac{T_S + t_k - T_G}{T_W}\right)^2 & , \quad -T_W \le t_k < 0 \\[2ex] 1 & , \quad 0 \le t_k < T_G \\[2ex] \left(\dfrac{T_S - t_k}{T_W}\right)^2 & , \quad T_G \le t_k < T_S \\[2ex] 0 & , \quad T_S \le t_k \end{cases} \qquad (6.3)
$$

and

$$
f^{DAB}_{unwanted} = \begin{cases} 1 & , \quad t_k < -T_W \\[2ex] 1 - \left(\dfrac{T_S + t_k - T_G}{T_W}\right)^2 & , \quad -T_W \le t_k < 0 \\[2ex] 0 & , \quad 0 \le t_k < T_G \\[2ex] 1 - \left(\dfrac{T_S - t_k}{T_W}\right)^2 & , \quad T_G \le t_k < T_S \\[2ex] 1 & , \quad T_S \le t_k \end{cases} \qquad (6.4)
$$

Figure 6.3 shows the curves corresponding to the definitions (6.3) and (6.4).

Clearly, other receiver synchronization strategies are possible, too. They are already implemented in receivers on the market. The positioning of the evaluation window by aligning to the first arriving signal can be a very bad choice indeed.

This might happen if, for example, the direct path between receiver and the nearest transmitter is blocked by an obstacle. Even if there

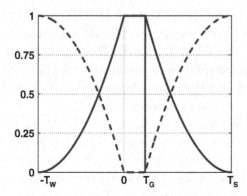

Figure 6.3: T-DAB weighting factors for wanted (solid) and unwanted (dashed) contributions of delayed signals.

is still some power arriving along that way, the attenuation suffered
by the presence of the obstacle might be so large that the main signal
contribution comes from a different transmitter site. If its corresponding
time delay is larger than the guard interval, severe degradation of the
reception quality up to total failure would result.

An obvious approach to exploit the arriving signal in a more efficient
way is to position the evaluation window according to the strongest contribution. However, if this happens to be a single signal that is separated
by a lap of time larger than the guard interval from a group of other signals then once again self-interference might occur.

The optimal strategy with respect to improving the reception quality
as much as possible certainly is to adjust the evaluation window such
that the ratio between useful and interfering contributions is maximal.
The drawback of this method is the need for more complex receivers
and corresponding sophisticated algorithms to find this position of the
evaluation window. More details on receiver synchronization strategies
can be found in [EBU03].

Whatever synchronization strategy is employed by the receiver, it is
evident that this is one of the features that differentiates good and bad
receivers. This is exactly the issue where the know-how and the experience of the manufacturer becomes relevant both in terms of performance
as well as costs of production of the receivers. Hence, it comes as no
surprise that there is nearly no information available about the actually
implemented synchronization strategies.

Simulations and hardware tests of DVB-T systems have shown that the receiver model appropriate for T-DAB according to (6.3) and (6.4) cannot be applied to DVB-T as well. The main reason lies in the fact that both systems use different modulation techniques. T-DAB employes a differential QPSK modulation scheme, while DVB-T rests on coherent QPSK or QAM modulations (see Chapter 2).

The coherent demodulation in the case of DVB-T requires an estimation of the channel response function. The pilot carriers are utilized thereto. Strictly speaking, the impulse response function then is only known for those frequencies the pilot carriers are located at. Therefore, a spectral interpolation is carried out based on an appropriate filtering process. The corresponding filter bandwidth leads to a characteristic time interval T_F that influences the receiver properties in the presence of interfering signal components [EBU03]. Echoes arriving later than T_F, thus, introduce degradations in the determination of the impulse response, and thus contribute only in a destructive manner.

As in the case of T-DAB, signals having a spread of their times of arrival smaller than the guard interval are considered to contribute constructively. Signals arriving later but before T_F are also split into wanted and unwanted parts, taking into account adequate weighting factors. All contributions with a time of arrival beyond the filter time T_F are considered to be interfering only. In mathematical terms the weighting functions can be written as

$$
f_{wanted}^{DVB} = \begin{cases}
0 & , \quad t_k < -T_F \\[2ex]
\left(\dfrac{T_S + t_k - T_G}{T_W} \right)^2 & , \quad -T_F + T_G \leq t_k < 0 \\[2ex]
1 & , \quad 0 \leq t_k < T_G \\[2ex]
\left(\dfrac{T_S - t_k}{T_W} \right)^2 & , \quad T_G \leq t_k < T_F \\[2ex]
0 & , \quad T_F \leq t_k
\end{cases}
\tag{6.5}
$$

and

$$
f^{DVB}_{unwanted} = \begin{cases}
1 & , \quad t_k < -T_F \\[2ex]
1 - \left(\dfrac{T_S + t_k - T_G}{T_W} \right)^2 & , \quad -T_F + T_G \le t_k < 0 \\[2ex]
0 & , \quad 0 \le t_k < T_G \\[2ex]
1 - \left(\dfrac{T_S - t_k}{T_W} \right)^2 & , \quad T_G \le t_k < T_F \\[2ex]
1 & , \quad T_F \le t_k
\end{cases} \quad . \; (6.6)
$$

Figure 6.4 shows the curves corresponding to the definitions (6.5) and (6.6).

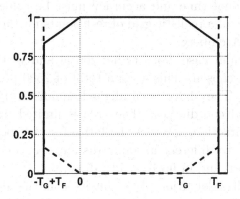

Figure 6.4: DVB-T weighting factors for wanted (solid) and unwanted (dashed) contributions of delayed signals.

6.1.4 Service Area

All existing wave propagation models can make statements only for single points and not for areas. Thus in practice, predicting the field strength throughout an extended area is approximated by predictions for a large number of individual points usually called test points. The more points are used, i.e. the smaller the gaps are between them the more accurate the results usually become. Quite often, it is necessary to vary the density of calculations points across an investigated area. This refers to situations in which for a small part of the original area a higher spatial resolution is required, because, for example, an urban area is included.

Another important issue in relation to the calculation of field strengths concerns the actual geographical location of the considered points of reception that represent the explored service area. In order to deal with a wide-area coverage like an entire national territory, an equidistant grid is a straightforward way to define the test-point distribution. The location of the test points then coincides with the grid points. Obviously, the number of test points that fall into a particular service area depend on the horizontal and vertical distances of the grid points.

The choice of these two lattice constants is critical with respect to the computational effort, when evaluating the coverage situation throughout the considered area. Too small lattice constants will lead to unacceptably high computer run times, whereas too large values will restrict the worthiness of the coverage prediction. Thus, a reasonable trade-off between computational time and accuracy must be achieved. Figure 6.5 shows an example of a realistic grid of test points for the German federal state Baden-Württemberg.

In this particular case, both in horizontal and vertical direction 50 points have been chosen. This gives a total of 2500 points to represent the service area. They have been aligned equidistantly with regard to their geographical coordinates. The area is limited by 7.5° and 10.5° in longitudinal direction, whereas the latitudes vary between 47.5° and 50.0°. This leads to different absolute distances between grid points in east–west direction and in north–south direction.

Clearly, another definition of the lattice could be employed as well. First of all, it is not necessary to use geographical coordinates to generate the grid. With the same right the geometrical distance between points could be fixed to, for example, 1 km. Furthermore, those test points lying outside the boundaries of Baden-Württemberg could be neglected. However, a rectangular structure as seen in Figure 6.5 allows to produce faster software due to simpler access to the set of test points.

Nevertheless, selecting the test points according to a grid neglects completely those different points within an area that can be of totally different importance for the network provider. Even a public broadcaster

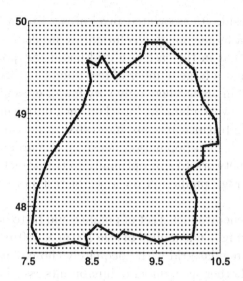

Figure 6.5: Example for a set of test points based on a rectangular
grid as basis for network planning in Baden-Württemberg, Germany.

whose primary obligation is to provide wide-area coverage might distinguish between test points being located in urban areas or in the middle
of a forest. The proximity of a test point to main traffic routes might be
relevant, too. Moreover, some times certain test points have to be taken
into consideration, because the topography in the vicinity has special
characteristics that seem to be important. Also, it might be necessary
to take into account certain test points because they have economic or
political relevance.

Apart from the major traffic routes, the population distribution certainly is one of the most important factors that can determine the selection of a set of test points. Employing a grid surely is very well adapted
to wide-area coverage problems. This is not valid for population density
oriented planning activities.

The statistical methods employed in connection with appropriate
wave propagation models refers to pixels of a size depending on the chosen resolution. A typical value for the resolution is a pixel representing
an area of 100 m × 100 m (see Chapter 4). Consequently, each test point
corresponds to a pixel as defined in the context of coverage prediction.
Thus, each test point can be endowed with the number of people that

happen to live inside its associated pixel area. To this end, population data bases with a large enough resolution are needed.

It is evident that in the case of a grid based test point generation it might turn out that a large number of test points are taken into account, which are located in areas where only very few or even no customers live. Obviously, any network planning based on such data will not be optimal. Especially private broadcasters are interested in providing services only in those areas with high-population densities like urban areas. In order to configure the network to suit optimally to this target, it is reasonable to base the coverage prediction on a set of test points that maps the coverage target.

One way to generate such a set of points could be to define a threshold of inhabitants living inside a pixel of size $100\,\mathrm{m} \times 100\,\mathrm{m}$. Each test point resulting from a regular grid with adequately chosen lattice constants is assessed to see whether the number of inhabitants associated to it exceeds the threshold or not. If the actual figure is larger than the threshold then the corresponding point is included in the set of test points, if not it is neglected.

Applied to the example of Figure 6.5, a threshold value of 30 inhabitants inside 100 m × 100 m pixel results in a more or less uniform distribution across the area of Baden-Württemberg. A corresponding example is shown in Figure 6.6. If the threshold is chosen significantly larger like for example, 300, then only points belonging to the highly populated urban areas in Baden-Württemberg are left. This can be seen in Figure 6.7.

6.2   Network Topologies

Terrestrial broadcasting aims to cover large areas. Depending on the frequency, the wave propagation conditions allow to cover entire countries or even beyond, with one single transmitter. Typically, this is the case for frequencies below 30 MHz. In the VHF regime, depending on the ERP of the transmitter the range is reduced to a coverage radius of 100 km and below. At 1.5 GHz feasible coverage areas for transmitters shrink down to less than 10 km. Therefore, it is obvious that in order to provide a broadcasting service throughout a country of the size of, for example, Germany or even one of its federal states as outlined in Figures 6.5–6.7, a set of transmitters has to be employed.

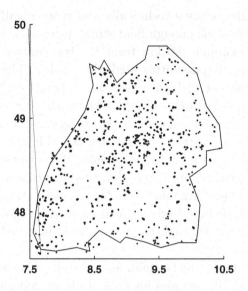

Figure 6.6: Example for a set of test points based on population data. Only points with a moderate number of inhabitants are accepted.

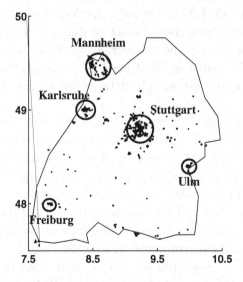

Figure 6.7: Example for a set of test points based on population data. Only points with a large number of inhabitants are accepted.

If at a given frequency a technically and economically feasible transmitter is able to provide enough field strength for good reception within a radius of, for example, 50 km from the transmitter site, then basically every 100 km a transmitter might be needed. This is just a rough guess for the time being, in practice this depends on topography and wave propagation conditions. At the edge of the service area of a chosen transmitter, signals from other transmitters will be received as well. Since analogue broadcasting systems are not able to cope with multipath environments as are COFDM systems, those transmitters that can be received at a given point of reception need to utilize different frequencies. Only after a very large distance, frequencies can be reused without causing unacceptable interference. Such a network topology used to provide a single content throughout a large area is called multifrequency network.

To give an example, the German federal state of Baden-Württemberg can be considered. It occupies an area of about 200 km × 250 km. In order to provide one single FM program throughout this topographically difficult region, a total of more than 10 high-power transmitters are needed, using different FM frequencies.

The situation is quite different for digital terrestrial broadcasting systems based on COFDM technology. As described in Section 2.1, multipath environments give rise to a set of different signal contributions, arriving at the location of the receiver. In contrast to analogue systems, this multitude of signals can be utilized in a constructive way up to certain limits. The limits are mainly determined by the magnitude of the guard interval.

If the delay spread between the first and the last arriving signal is larger than the guard interval, self-interference will be encountered. Nevertheless, the COFDM technology allows to build SFNs, which means that the same content is provided throughout a large area such as that of Baden-Württemberg using the same frequency, that is a T-DAB block in the case of T-DAB or a TV channel for DVB-T. In the first place, this means that SFN technology allows to significantly reduce the amount of required spectrum for a given coverage target.

Furthermore, in geographical regions with a lot of mountains and steep valleys SFNs offer even more advantages. If the direct path between transmitter and receiver is blocked by some obstacle, such as a mountain, then it is still possible that a satisfying reception quality can be achieved with the help of reflections.

In the case of large area broadcasting coverage, both MFN and SFN topologies are usually based on a limited set of high-power transmitters. Actually, this a very efficient way to provide a broadcasting service throughout a large area, in case a reception mode is envisaged that does not require high-minimum field strength value to guarantee a satisfying service. However, in particular, if handheld reception or portable indoor reception in big cities with large concrete buildings is intended, then significantly higher field strength values are required. In that case, it might not be possible to increase the ERP of the high-power stations correspondingly. Apart from technical constraints also public health issues might become important, which impose to obey certain field strength levels not to be exceeded. Additional transmitter sites are then necessary.

Depending on the circumstances a very large number of transmitters concentrated in a rather small area might be utilized. The network is then called a dense network. Clearly, in that case it is no longer necessary to use high-power stations. Rather, instead of employing 100 kW transmitters every 100 km it is possible to work with 1 kW or less at each transmitter site. It has to be noted, however, that the dense network option is neither always reasonable nor always technically feasible. For a wide-area coverage the number of needed transmitters increases to several hundreds instead of 10–15. Even though a small transmitter is significantly cheaper concerning both acquisition costs and operational costs, the signal feeding from the play-out center to the transmitter site might impose a challenge both in terms of technical realization and associated costs. Therefore, a network provider needs to very carefully analyze, which type of network topology is the most effective and cost efficient in order to achieve given coverage objectives.

6.3 Network Optimization

Any transmitter network is build from a set of N transmitters that are described by defining N sets of network parameters as listed in Table 6.1. In order to assess the quality of this network, it is necessary to specify a wave propagation model, a receiver model, and to define the envisaged service area.

A typical planning scenario for a digital terrestrial single frequency network consists of about 10–30 transmitter. However, a significantly larger number might be necessary in some cases. Planning the network means, to adjust the set of network parameters in order to achieve a defined quality criterion.

From a mathematical point of view the parameters ERP, time delay, and antenna factors are real parameters, that is they can take arbitrary real values. However, for practical reasons, it is reasonable to introduce a set of distinct values from which each of them has to be chosen, respectively. If, for example, each of the network parameters is associated to only 10 different values to choose from then already for one single transmitter there are $10 \times 10 \times 10 \times 10^{36} = 10^{39}$ different configuration possibilities. Considering a small network of only 10 transmitters, therefore gives rise to the astronomically large number $\left(10^{39}\right)^{10} = 10^{390}$ of different network configurations. Practically, this is equivalent to infinity.

As in the case of frequency assignment problems, obviously also here, a NP-hard combinatorial optimization problem is encountered. Any brute-force approach based on pure trial-and-error must necessarily fail. Once again, it seems that more sophisticated methods are needed in order to find a satisfying configuration of the network. Stochastic optimization algorithms are once again promising candidates.

However, it turns out that employing the network parameters of Table 6.1 leads to combinatorial problems that are very, very complex. All stochastic optimization algorithms evaluate only a very small fraction of the total number of possible configurations. The determination of a satisfying solution is mainly connected to the application of more or less sophisticated strategies to decide, which configurations are tested and which not.

The large number of allowed values for the network parameters is connected to the commonly used description of antenna patterns in terms of 36 different values. Such a parameterization does not make too much sense when seen from the optimization algorithm's point of view. Hence, it seems to be reasonable to introduce a representation of an antenna diagram, which is based on a significantly smaller number of parameters.

Basically, there are two promising possibilities. Looking at real transmitters indicates that both the shape as well as the direction of an antenna diagram is usually subject to restrictions. Therefore, it would be natural to define 3–5 different antenna diagrams that can be employed. Furthermore, it could be assumed that these diagrams could be directed only towards 36 directions, namely from 0° to 350° with a step-size of

$10°$. Instead of 10^{36} there would be only $5 \times 36 = 180$ possible antenna configurations for one transmitter.

Another possibility would be to define antenna diagrams in terms of mathematical formulas, depending on a small number of free parameters only. A very simple diagram can be described by two parameters, for example. One of them is the angle Φ_k of the direction of the main lobe. The other specifies the width κ_k of the antenna diagram.

It is assumed that the vector

$$\mathbf{s}_k = \begin{pmatrix} \cos \Phi_k \\ \sin \Phi_k \end{pmatrix} \tag{6.7}$$

is oriented along the direction of the main lobe while

$$\mathbf{d}_k = \begin{pmatrix} x - x_k \\ y - y_k \end{pmatrix} \tag{6.8}$$

represents the line connecting the location of transmitter k and the point of reception. Calculating the scalar product between $\mathbf{s}_k$ and $\mathbf{d}_k$ allows to determine the angle ϕ between the two vectors in Equation (6.7) and (6.8):

$$\phi = \arccos \left\{ \frac{(x - x_k) \cos \Phi_k + (y - y_k) \sin \Phi_k}{\left[(x - x_k)^2 + (y - y_k)^2 \right]^{1/2}} \right\} . \tag{6.9}$$

It has to be noted that the definition (6.9) implies the angle zero to coincide with the east direction, which is in contrast to the common geographical definition where the north direction is usually considered as representing zero degrees. Furthermore, the angle increases by counterclockwise rotation.

With the help of (6.9) a simple antenna diagram can be defined as

$$A(\phi, \Phi_k, \kappa_k) = \exp \left[-\kappa_k \phi^2 \right] . \tag{6.10}$$

Applying this antenna pattern means the ERP has to be multiplied by $A(\phi, \Phi_k, \kappa_k)$ in order to calculate the field strength in the direction from the transmitter to the point of reception. In logarithmic representation the multiplication corresponds to adding the term

$$10 \log_{10} \left[A(\phi, \Phi_k, \kappa_k) \right] = -10 \, \kappa_k \phi^2 \log e \tag{6.11}$$

to the ERP [Beu95]. If both for the parameter κ_k as well as for Φ_k 10 different values are allowed, a total of $10 \times 10 = 100$ configurations is possible.

The function defined in Equation (6.10) is monotonous, and hence has only one local maximum. The corresponding antenna diagram thus shows exactly only one main lobe. It has to be checked on a case-by-case basis, if such an approach is reasonable or not. By superimposing several equivalent terms corresponding to different values κ_k and Φ_k it is in principle possible to create a diagram having more significant lobes. The price to pay, however, is a corresponding increase of the number of network parameters.

6.3.1 Assessment of a Network Configuration

Fixing the set of network parameters to certain values corresponds to establishing a particular network configuration. Clearly, an arbitrary choice can give rise to very poor coverage. Therefore, it is necessary to select from the vast number of possible configurations one that gives satisfying coverage results.

However, the terms "good" or "bad" in relation to a network configuration corresponds necessarily to a relative assessment of the network. It crucially depends on the way the quality of a network is defined. What represents an optimal choice with respect to a particular quality function might be considered bad in the context of other quality criteria.

Furthermore, coverage obviously is a two-dimensional concept. Therefore, it might turn out that globally a certain network configuration is being considered good. At the same time, however, there might be very unsatisfying coverage conditions on a local level in special regions of the service area. Clearly, the inverse situation might be encountered as well. This phenomenon has to be taken into account properly when defining the quality function by which the network is to be assessed.

If wide-area coverage is the primary coverage target, it is obvious to select a sufficiently large number of test points spread out across the envisaged service area and use them as a representative base for an appropriately chosen quality measure. In general, there are three types of test points that are relevant for an assessment of the network configuration of a SFN. There are served points, points where the minimum field strength is not reached, and points where no satisfying service can be provided due to self-interference problems.

Quite generally speaking, one way to cover a larger area is to simply increase the emitted power in the network. However, more power output entails significantly higher costs. In addition, for digital single frequency networks increasing the power of individual stations in the network might lead to an increase of self-interference in first place rather than increasing the service area.

A natural starting point to find a good network configuration is to adjust the network parameters such that the number of served test points Z_S is maximized. If this does not lead to satisfying results it is worthwhile to try to minimize the number of points with self-interference Z_{SI} and the number of points where the minimum field strength Z_{min} is not reached. Very often a combination like

$$Q = \frac{Z_S}{(1 + Z_{SI})(1 + Z_{min})} \tag{6.12}$$

of all three figures gives the satisfying results. In order to apply the quality criterion (6.12) the total number of test points,

$$Z_T = Z_S + Z_{SI} + Z_{min} , \tag{6.13}$$

must be kept fixed.

Equation (6.12) can be used as quality function, for example, when employing a stochastic optimization algorithms. Then, the task is to find that set of network parameters which corresponds to a maximum value of the quality Q. In Equation (6.12) the denominator has been chosen such as to avoid the occurrence of singularities. Other combinations of the three figures Z_S, Z_{SI}, and Z_{min} can be utilized as well.

By using nonlinearities like exponential functions it is possible to introduce different weights. There is actually no limit with regard to inventing new quality functions. Sometimes it is even unavoidable to put some effort into the creation of a reasonable quality function because it might turn out that a straightforward definition like such as Equation (6.12) is not sensitive enough.

Apart from the different types of test points it might be necessary to take into account the number of served inhabitants, too. Each test point is associated with a number of people. Hence, the summation of the number of served test points can be supplemented by a summation of the corresponding number of inhabitants E_k associated with the pixel area of the test point. In the same way, the operating costs K_k for the transmitters can be taken in account.

If the coverage target is to maximize the number of served test points and the number of served inhabitants along with a simultaneous minimization of the number of interfered or unserved pixels and the network costs, then a quality function like

$$Q = \frac{A\, Z_V + B\sum_{k=1}^{Z_S} E_k}{C\,(1 + Z_{ESI})\,(1 + Z_{min}) + D\sum_{k=1}^{N} K_k} \qquad (6.14)$$

could be employed. In definition (6.14) the four scaling factors A, B, C and D have been introduced. By appropriately adjusting their values the optimization criteria gain or loose weight with respect to the application of the quality given by Equation (6.14). Whether such an approach will give satisfying results must be decided on a case-by-case basis.

As a matter of fact network optimization is a rather complicated business, having to deal with a lot of different constraints at the same time. Therefore, as in the case of constrained frequency assignment problems optimization algorithms are required, which provide enough degrees of freedom to take all constraints into account. Hence, it comes as no surprise that algorithms as discussed in Annex A are very often chosen.

6.4 Implementation of an Allotment Plan Entry

The features of digital terrestrial broadcasting systems show to advantage, if they are operated as SFNs. However, a successful operation presupposes that the two aspects "frequency management" and "network planning" are not considered independently from each other.

International radio conferences are the forum where frequency resources are assigned to submitted requirements according to agreed principles (see Chapter 7). Administrations asking for frequencies in order to provide services such as T-DAB or DVB-T, have to very carefully define their requirements such that apart from any media political or economical constraints a reasonable network operation is feasible from a technical point of view.

Obviously, defining allotment areas as large as possible favors efficient usage of spectrum. However, very large service areas meant to be served in terms of SFNs run the risk of having to cope with self-interference. In principle, this can be compensated by careful tuning of the network parameters. It should be noted, however, that in practical terms this is not always feasible due to economic or technical constraints.

Furthermore, it has to be emphasized that not the absolute extension of allotment areas is the relevant parameter for efficient spectrum usage, but rather the ratio between the minimal extension of areas and the reuse distance determines the spectrum demand. If this ratio is larger than one, only the directly flanking areas of a particular allotment have to be considered when looking for a frequency that is not in conflict with others. In that case, next but one neighbor areas are not relevant. This can be deduced from the results of Section 5.7.1. If the ratio between typical minimal extension of an allotment and the relevant reuse distance drops below one, additional spectrum is required.

In any case, it is crucial for network providers to carry out extensive studies concerning the size of allotment areas before they request frequencies for their requirements. Usually, such investigations are based on existing transmitter sites. Topographical conditions have to be taken into account in order to come to a realistic assessment of the coverage situation.

Nevertheless, any frequency plan generation on an international level will be based on rather simple assumptions in order to obtain a very general result. This refers in the first place to the application of wave propagation models like ITU Recommendation ITU-R P.1546 [ITU01a]. Moreover, for realistic transmitter networks, simple models are employed like for example, the reference models described in Section 5.3. Consequently, the resulting frequency plan gives a theoretical framework for the usage of frequencies under which real network implementations have to be carried out.

Network implementation is primarily driven by the wish to provide a certain service throughout a given area in an economic and efficient way. The number of transmitters, their location, and their technical characteristics need to be carefully adjusted to achieve the intended coverage objectives. A network provider will always need to weigh up the increase in service area or service quality against the network implementation efforts in terms of money and resources.

But network implementation cannot only focus on obtaining a certain coverage objective inside a given area. Since any network implementation has to make reference to some entry in an international frequency plan it is obvious that the protection needs of other existing transmitter networks or plan entries not yet brought into operation have to be taken into account. This imposes constraints on the technical characteristics of the transmitters employed. In particular, the ERPs and antenna diagrams of transmitters have to be regulated in a way to reduce the interference into those geographical directions where other networks are in operation or could be implemented according to corresponding frequency plan entries.

Usually, network implementation starts with the selection of transmitter sites. This is mainly determined by the fact that transmitter sites for broadcasting purposes cannot be easily erected wherever it might seem to be appropriate. So, in most cases broadcasting network providers have to stick to a set of existing sites. Among them, there are usually main stations capable to emit large power while others can be operated with limited power only. Furthermore, most transmitter sites are not all-purpose sites. This means, in addition to a power restriction there are often other constraints concerning, for example, the degrees of freedom to adjust an antenna diagram according to what might be necessary. At a given site mast statics could hinder the addition of further antennas. This could basically mean that a site cannot be used even though its geographical location might be advantageous.

Having selected a set of promising transmitters sites, the network planners start to carry out calculations to predict the service quality assuming realistic transmitter characteristics. However, as mentioned above, not only the achievable service area has to be accounted for, but also the interference that will be produced by the planned network at given geographical locations. These locations are clearly defined in connection to the underlying frequency plan and usually they correspond to either points along the national border of a country or the border of a service area of another broadcasting service that is to be protected.

In practice, this results in a large number of different constraints that have to be met in order to design the network properly. Therefore, the whole planning process is an iterative process that traditionally has been carried out on a trial-and-error basis. A first set of transmitters is chosen, technical characteristics are defined, and a service prediction is carried out. At the same time the interference impact on other networks is evaluated. Since the first trial will certainly not lead to satisfying results, the network implementation is modified and the calculation process is repeated until a satisfying solution is found.

Iterative processes to find a solution to a problem are perfect candidates for stochastic optimization. This is also true in the case of network implementation of allotment entries of frequency plans. The problem is high-dimensional, that is there is a large number of parameters that have to be adjusted, and the space of potential solutions is incredibly large. Furthermore, there is a vast number of very different constraints that need to be considered, too. Basically, the same approach as described in Section 6.3 can be applied. The only difference is that the compatibility with respect to other existing networks or plan entries has to be taken into account in addition.

As already mentioned, one of the most important issues concerning real network implementation is the selection of the transmitter sites. The question is how this task can be tackled by means of stochastic optimization modeled as in Section 6.3. The approach discussed there presumes that the number of transmitters and their geographical location is fixed. This is not an important assumption, because in principle it is possible to treat the location of the transmitters as planning parameters as well.

There are, however, two reasons why this is usually not done. The first one is that from a network providers point of view there is not so much freedom in choosing the sites of the stations, because there is only a limited number of appropriate sites in any case and there is no possibility to make accessible further transmitters.

Second, computational time becomes an issue, if site selection is included in the network planning process. The strategy outlined in Section 6.3 requires to calculate a full service prediction for each set of temporarily chosen network parameters. In order to accelerate the computations, a set of properly chosen calculation points is used. For each of them a normalized field strength value is calculated for each of the transmitters. Applying the current network parameters to the transmitters, then simply means to scale the field strength, correspondingly. If the location of a transmitter can be freely chosen, then each iteration of the stochastic optimization requires to carry out a full wave propagation prediction.

Nevertheless, it is possible to introduce site selection into the network planning process. To this end, a list containing all transmitters that could be used for the network is set up. Each transmitter is defined by its individual set of allowed network parameters such as ERP, antenna diagram, and so forth. Site selection can then be simply introduced by allowing a transmitter to get an ERP of 0 kW. Hence, zero out power is included in the list of possible ERPs for each transmitter. Therefore, if at any time the simulation assigns a zero power to a transmitter, it is basically not part of the network. At the end of optimization simulation, only those transmitters that are actually needed for the achievement of the coverage objective will be assigned an ERP greater than zero.

For the implementation of an allotment plan entry, this approach is also advantageous in relation to meeting the acceptable interference levels imposed on other networks. In addition to the set of calculation points inside the allotment area, another set of calculation points is defined, where the total interference of the network is evaluated. The objective function has to be modified in a way to favor such network configurations that produce acceptable interference levels at all these points. As a consequence, solutions for the network implementation might be found, which discard transmitters close to the boundaries of other networks or other allotment plan entries, that is they will be assigned zero ERP. At the same time the ERP of transmitters inside the allotment area under consideration will be increased in comparison to a planning exercise where outgoing interference is not an issue.

Chapter 7

The Regional Radiocommunication Conference RRC-06 and the GE06 Agreement

The usage of the electromagnetic spectrum is managed by the international organizations and bodies described in Chapter 3. Whenever a new type of broadcasting system is introduced or a part of the spectrum that has been used by other systems is to be allocated to broadcasting, it is necessary to establish a corresponding detailed frequency plan. Such a plan contains a list of transmitters or equivalent planning objects together with a specification of their technical characteristics and the frequency that can actually be used. Moreover, a frequency plan needs to be amended with a technical and regulatory framework, providing the means to modify or extend the plan. Typically, frequency plans are addressed as "arrangements" or "agreements" between administrations of a certain region of the planet, which signed the corresponding final acts of a planning conference.

Several frequency plans for analogue broadcasting have been set up over the last 100 years. The most important for Europe relating to terrestrial television was the so-called Stockholm Plan (ST61) established on ITU level by a corresponding conference that was held in Stockholm, Sweden, in 1961 [ITU61]. It governed the usage of spectrum for terrestrial television for more than 40 years. For Africa and the Arabic

R. Beutler, *Digital Terrestrial Broadcasting Networks*,
DOI 10.1007/978-0-387-09635-3_7, © Springer Science+Business Media, LLC 2008

countries, a corresponding frequency plan has been set up in Geneva, Switzerland, in 1989 (GE89) [ITU89]. With the advent of digital terrestrial broadcasting, it became necessary to develop a new plan for the new digital systems. This process culminated in the GE06 Agreement, which contains a frequency plan for DVB-T and T-DAB in the VHF and UHF range [ITU06]. The new plan superseded ST61 in Europe and GE89 in the African Broadcasting Area (ABA).

The introduction of digital terrestrial broadcasting in Europe started more than 10 years ago. As this was not an ad hoc process several phases had to be passed through and there were different needs at different times. The first system for which a new frequency plan was needed was T-DAB. Later digital television was considered as well. This chapter gives an overview of those arrangements and agreements that constituted the basis for GE06.

7.1   Wiesbaden 1995

In July 1995, CEPT organized a planning meeting for T-DAB [CEP95] that was held in Wiesbaden, Germany (WI95). Administrations were called to submit their requirements to that conference whose task was to provide appropriate frequencies for them. The planning conference succeeded in drawing up a frequency plan for T-DAB providing two nationwide coverages for each of the 43 participating CEPT member countries. They signed the Final Acts of the "CEPT T-DAB Planning Meeting, Wiesbaden, 1995." A detailed overview can be found in [Hun96] and [OLe98].

WI95 was the first planning conference in Europe, dealing with digital terrestrial broadcasting systems. Therefore, there was basically no source from which the planners could extract planning methods and algorithms, in order to assemble a frequency plan for T-DAB. As a consequence, a very extensive preparatory phase preceded the conference. Both EBU and CEPT were heavily involved in the preparations. Most of the concepts that nowadays are common ground have been invented and developed in relation to WI95.

The purpose of WI95 was to establish a frequency plan that would allow to offer two multiplexes of T-DAB throughout the national territory of an administration. The major problem was the availability of an appropriate amount of spectrum across the CEPT planning area.

It was agreed to use frequencies from several bands for the T-DAB plan, namely the spectrum ranges 47–68 MHz (Band I), 174–230 MHz (Band III), 230–240 MHz and 1452–1467.5 MHz (L-Band).

Band III was heavily used by analogue terrestrial television and also by other primary services, such as fixed links or aeronautical telemetry at that time. The only resource that could be made available almost everywhere in Europe was VHF channel 12. Therefore, the first T-DAB layer was realized by using channel 12 as much as possible. In some cases, parts of the planning area channel 13 has been used as well, for example, in Scandinavia. On a regional level also channels below 12, that is VHF channels 5–11, have been employed. The second layer was established with the help of T-DAB blocks accommodated in the L-Band. However, there were exceptions from these rules. In United Kingdom, L-Band was not available for broadcasting at that time, therefore the UK layers were both accommodated in VHF. In France, it was just the other way round. VHF was not available for T-DAB, and so both the French T-DAB layers were generated with the help of L-Band frequency blocks.

One of the novelties of WI95 was the introduction of allotment planning for terrestrial broadcasting. All frequency plans for terrestrial analogue systems before were based on pure assignment planning. This means entries to a frequency plan were given in terms of the technical characteristics of the corresponding transmitters sites. In contrast, allotment planning employs an abstract object for planning purposes. An allotment that is in the first place defined by its area acts as a placeholder for real transmitters (see Section 5.1).

The decision to adopt allotment planning was based on two arguments. First of all, most of the countries did not have detailed network plans available for T-DAB at the time of the conference. Therefore, the planning for T-DAB had to be flexible. Second, the technical characteristics of T-DAB being a COFDM system allow to operate single frequency networks (SFN) (see chapter 6). This means a T-DAB network can be developed and extended gradually and subsequently adapted to different coverage needs without problems, if operated in SFN mode. New transmitters can be added to the SFN when necessary or their technical characteristics can be modified, if required. Clearly, this is of enormous benefit for the network provider and this feature should be incorporated into a corresponding frequency plan.

If planning is built on allotments it is necessary to define both their protection needs and interference potential in detail. This was the reason why the concept of reference networks (RN) has been invented. Due to the differing propagation conditions between VHF and L-Band two separate RNs have been established, one for each spectrum range. The technical details of both RNs can be found in Section 5.3.

The plan generation process at Wiesbaden was guided by the general principle of equitable access to spectrum. This was and still is considered as a very important general principle, which is applied in all international planning conferences. Even though it can be intuitively understood what is meant by that, it is very difficult to define it properly, that is quantitatively within a given planning context. Basically, equitable access means that all nations have the same rights to utilize any telecommunication service. Irrespective of political bias or economical power, any country can claim equal demand for radio or television services. This is also independent of the geographical location or the size of a country. The same holds for the number of inhabitants or their geographic distribution across the area of the considered country. At WI95 this issue was solved by planning two nationwide T-DAB coverages for each country. It was left to the administrations to decide whether they wanted to provide nationwide or regional services inside their territories.

At the very beginning of the preparation of WI95, there was still dissent whether planning should be computer aided or not. In the end, EBU and ERO had developed planning tools that were finally used. They have been employed ever since for all subsequent broadcasting planning conferences. Clearly, their details have been modified in order to fulfill the actual planning tasks, respectively.

The basic planning principle of the planning process consisted of two independent steps, namely compatibility analysis and plan synthesis. These two steps are meant to be applied several times in an iterative manner until a frequency plan can be agreed upon by all involved parties.

Administrations provided their input requirements to the planning process in electronic form. At a first step, they underwent an extensive analysis to identify incompatibilities between pairs of T-DAB requirements, T-DAB requirements and existing analogue TV stations, and finally between T-DAB and any other existing primary services. The result of this compatibility analysis indicated, which requirements could share a T-DAB block. In addition, during this process a list of available frequencies was provided for each T-DAB requirement.

During the preparation of WI95 the idea was expressed that administrations could examine these results and permit adjustments to be made,

such that additional frequencies became available for particular requirements. However, in practice this step was not implemented in that way
at the conference.

After the input data sets had been prepared, the plan synthesis process was started. It was based on graph theoretical algorithms like the
ones described in Section 5.7.2. A detailed presentation of the calculation method can be found in [OLe98]. The plan synthesis resulted in a
frequency allocation for a part of the requirements. Due to incompatibilities or the large number of requirements struggling for T-DAB blocks
in particular regions, there were many requirements that could not get
a frequency.

Specifically for these cases, administrations tried to reach agreement
about sharing of frequencies. The basis for these corrections were interference calculations, using more sophisticated wave propagation models
in particular taking into account topography and morphology. The application of these more advanced methods showed that in many cases
less interference could be expected than predicted by the calculations
of the compatibility analysis. However, also the inverse situation was
encountered. Therefore, higher mutual interference levels were agreed
between neighboring countries in a large number of cases, as well. The
resulting agreements have been taken into account during the subsequent
iterations of compatibility analysis and plan synthesis. In the end, an
allotment plan was drawn up, which accommodated about 750 T-DAB
requirements in VHF and L-Band, in many cases subject to special constraints for future network implementations.[8]

WI95 was convened under the provisions of the ITU Constitution and
the Radio Regulations in particular Article 42 of the Constitution and
Article 7 of the RR. Basically, this means that neither the conference nor
its results should be in conflict with the general ITU framework and the
results of the conference shall be notified to the Secretary-General of the
ITU. The CEPT Special Arrangement, WI95 contains an allotment plan
for T-DAB together with procedures that are to be applied in order to
modify the plan or to implement allotment plan entries in terms of real
transmitter networks. The overall formal structure of the arrangement
text was derived from existing ITU plans like ST61. From a practical

---

[8]The 750 T-DAB requirements only made up for the smaller part of the total of
allotments and assignments that had to be dealt with WI95. About 40,000 other
primary services had to be taken into account during the compatibility analysis.

point of view, the most important articles of WI95 are the Articles 4 and 6 covering coordination with affected administrations in case of a plan modification (Article 4) and provisions in order to convert an allotment into a set of assignments (Article 6). The later is just another wording for implementing a plan entry.

WI95 was unable to complete its work in two respects, namely to fully specify the technical details to be taken into consideration when converting an allotment into a set of assignments, that is applying the Article 6 procedure. Furthermore, the procedures contained in Article 4 could not be finalized as well. An interim procedure was included at WI95 in order to allow coordination between administrations just after the end of WI95. However, this interim procedure worked on the basis of worst case interference scenarios. Therefore, one year later another CEPT meeting was held at Bonn, Germany, which finalized the work [CEP96] relating to both Articles 4 and 6. This planning meeting is referred to as the "CEPT T-DAB Planning Meeting (2), Bonn, 1996."

Before the Regional Radiocommunication Conference RRC-06 established the GE06 Agreement superseding WI95 in relation to T-DAB in Band III, the WI95 plan had been extended by roughly 200 additional plan entries, which were coordinated after WI95 on the basis of Article 4 procedures. More details concerning the current status of T-DAB networks in Europe can be found on the Website of ERO [ERO07a].

7.2   Chester 1997

Nine years before the RRC-06, CEPT laid the foundations for the planning of DVB-T in Europe. In Chester, United Kingdom, a CEPT meeting was held aiming to pave the way for the introduction of digital terrestrial television [CEP97]. The Chester meeting was intended to achieve a Multilateral Coordination Arrangement (CH97) relating to technical criteria, coordination principles, and procedures for the introduction of DVB-T. However, in contrast to WI95 no new frequency plan was generated and consequently none was attached to CH97. Rather, it was decided that the existing ST61 Agreement and the associated frequency plan should form the basis on which the introduction of DVB-T is to be accomplished. CH97 had to be considered as some kind of a supplement to ST61 for digital terrestrial television.

Both frequency bands, which are covered by ST61, namely 174–230 MHz, that is Band III and the Bands IV and V from 474–862 MHz were considered in CH97. All decisions relating to technical criteria contained in CH97 were based on the original ETSI standard [ETS97b]. As in the case of the Wiesbaden Arrangement WI95, cooperation between administrations represented by CEPT and ERO and EBU was the key to success. A total of 35 countries signed the Multilateral Coordination Agreement.

As a matter of fact, the spectrum envisaged for the introduction of DVB-T in Bands III, IV, and V was almost fully occupied by analogue terrestrial television. In addition, there were also a lot of other primary services, such as fixed links or aeronautical navigation, which made use of VHF or UHF spectrum. Consequently, almost the only possibility to start DVB-T transmission was the conversion of analogue to digital terrestrial television. Clearly, this conversion was subject to the influence of an analogue environment. Therefore, CH97 needed to provide the measures how analogue stations can be protected from DVB-T. Basically, the idea was to define a reference situation for analogue television, which meant to calculate the service area of analogue stations in an all analogue environment. DVB-T transmissions were required to protect these service areas. These kinds of investigations became a part of corresponding procedures of CH97, which governed the coordination between administrations. In a first guess, reducing the ERP of a station by 7 dB while converting it from analogue to digital transmission is a good starting point for successful coordination of DVB-T.

In the years after CH97, DVB-T has been put into operation in several countries. In addition, many test transmissions have been started as well. By May 2002, a reference situation for analogue television had been achieved after lengthy bi- and multilateral coordination. More than 88,000 analogue stations were included. However, it became obvious that a planning conference for DVB-T would be needed because the provisions of CH97 lead to a suboptimal utilization of spectrum. CH97 constitutes an extension of ST61 in order to allow the introduction of DVB-T under the ST61 roof. ST61 has been built and modified by applying the planning principles of analogue television. Therefore, it is obvious that using such a framework cannot be very well adapted to the different reception modes and planning parameters of DVB-T. Hence, the process to convene a planning process was initiated on ITU level and came to an end with the signing of the GE06 Agreement in 2006.

7.3   Maastricht 2002

The Wiesbaden Conference WI95 provided the basis for the deployment
of T-DAB services in Europe in an equitable way [Hun02]. At that time it
was generally accepted that the data capacity of two T-DAB multiplexes
at every geographical point would be sufficient to satisfy the demand.
However, very soon it became apparent that an additional layer would be
needed. In particular, there was a need for rather small allotment areas
in order to be able to provide T-DAB service even on a local level. In
case this would not be intended by administrations, there would always
be the possibility to provide the same content throughout a group of
adjacent allotment areas using different T-DAB blocks in different areas.

In 1997, the preparations for a new planning conference for T-DAB
were started within CEPT. This process was actively supported by EBU.
The first problem to be solved consisted in finding the spectrum in which
the additional T-DAB layer could be accommodated. It was a quite
natural solution to extend the spectrum range in L-Band for T-DAB
from 1467.5 MHz up to 1479.5 MHz thereby making available seven
more T-DAB blocks. Originally, this part of the spectrum up to 1492
MHz had been allocated to S-DAB and not surprisingly, the S-DAB
community was not very keen in giving away these frequencies for other
usages. However, due to the fact that S-DAB still was a concept rather
than a system in operation, CEPT members concluded that there is no
justification not to devote this spectrum to T-DAB.

Part of the conference preparations was concentrating on the question whether the technical planning parameters for T-DAB need to be
changed. However, it was concluded that this is not necessary. So, the
same parameter values were used as in WI95, apart from two changes or
extensions. The receiver noise figure of a T-DAB receiver was updated
according to information from the manufactures and an additional new
reference network was introduced. The latter reflected the intention to
give administrations the possibility to include local broadcasting services
as well.

The task of the new conference was to plan a third T-DAB layer based
on basically two conditions. First, requirements could be satisfied by allocating one of the newly available blocks between 1467.5–1479.5 MHz.
However, also the old WI95 T-DAB blocks could be used if such a choice
is not in conflict with other usages of the spectrum. Second, other primary services including S-DAB in the frequency range above 1479.5 MHz
had to be protected.

The same procedure to draw up a frequency plan as in WI95 was applied. This means the plan was set up iteratively by passing through a compatibility analysis and a plan synthesis, respectively. During the compatibility analysis it was checked if two requirements could share a frequency. In addition, those T-DAB blocks that were not available for individual requirements due to conflicts with existing WI95 allotments, S-DAB, or other primary services were identified for each input requirement.

The conference was finally held from June 10–18, 2002 in Maastricht, Netherlands. About 150 delegates participated and 33 administrations signed the Final Acts of the "CEPT T-DAB Planning Meeting (4), Maastricht, 2002" (MA02).[9] Roughly 1900 new allotments have been included in the frequency plan during the Maastricht meeting. Together with the L-Band allotments, stemming from WI95, a total of 2400 plan entries constituted the MA02 L-Band plan in the end. As in WI95 many of these could only get into the plan on the basis of more than 400 agreements signed between administrations.

7.4 Regional Radiocommunication Conference RRC-04

The CEPT meeting in Chester CH97 made it very clear that a new frequency plan for digital terrestrial broadcasting was needed. Therefore, in 2000, the CEPT administrations led by Germany requested the ITU to consider convening a Regional Radiocommunication Conference (RRC) for the revision of the Stockholm Agreement ST61. The main target was to pave the way for a rapid introduction of DVB-T in the European Broadcasting Area (EBA).

In 2001, the ITU Council agreed on Resolution 1185 aiming to convene such a RRC in the EBA. The frequency bands 174–230 MHz and 470–862 MHz were to be replanned for DVB-T and T-DAB in terms of two conference sessions. Member States of the ITU from the planning area of the regional agreement for VHF/UHF television broadcasting

---

[9]Actually, at Maastricht, CEPT separated the T-DAB plan in VHF from the L-Band plan. Therefore, two Final Acts were issued. The first one contained just the remaining and updated VHF plan together with the corresponding Arrangement text, that is the Articles and procedures. This CEPT Arrangement is referred to as the "CEPT T-DAB Planning Meeting (3), Maastricht, 2002." The L-Band allotment stemming form WI95 were tranferred to the new L-Band Arrangement [CEP02].

(GE89) [ITU89] in the African Broadcasting Area (ABA) and neighboring countries also expressed the wish to take part in such a RRC for digital terrestrial broadcasting. Hence, the decision was taken to convene a RRC for the "planning of the digital terrestrial broadcasting service in Region 1 (parts of Region 1 to the west of meridian 170°E and to the north of parallel 40°S) and in the Islamic Republic of Iran, in the bands 174–230 MHz and 470–862 MHz, in two sessions." Figure 7.1 sketches the planning area.[10]

The first session, which is usually called RRC-04 was to lay the technical foundations in terms of setting up and agreeing on planning parameters and criteria, while the second part, referred to as RRC-06, was to draft a new agreement that together with a corresponding new frequency plan should be adopted by the conference. In order to prepare for these two conferences, ITU decided to establish Task Group 6/8 (TG 6/8) to draw up a report for RRC-04. This report was to contain all relevant technical information that RRC-04 was invited to consider for integration

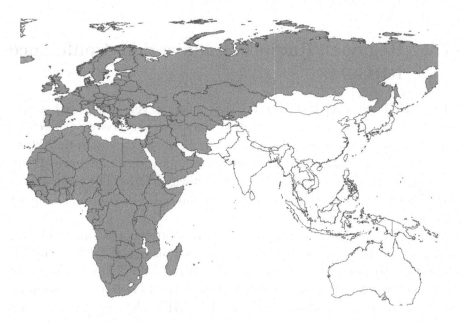

Figure 7.1: Planning area of the RCC-06. It extends up to 170°E.

---

[10]This figure has been produced with ArcGIS based on the world map contained therein.

into the new agreement. It covered planning concepts, technical criteria, and compatibility issues between different types of services and planning tools. This work was heavily supported both by CEPT and by EBU, and both set up corresponding project teams dealing with these issues. The report was eventually submitted to the RRC-04 together with further independent contributions from CEPT and EBU.

On May 10, 2004 the RRC-04 started in Geneva [ITU04a]. About 750 delegates from 95 countries participated in the conference lasting for 3 weeks. In contrast to other conferences in the past, the heads of delegation could not agree on a single candidate as chairman of the conference. Therefore, it was agreed that two "acting chairmen" should alternate as chairman. In the end, it turned out that this somewhat strange arrangement worked quite well. The structure of the conference itself was clear from the very start. Six Committees were created out of which, Committee 4 dealing with technical issues and Committee 5 concerned about planning issues essentially did the work.

RRC-04 agreed on planning parameters and criteria such as minimum field strength values, protection ratios, and so forth. However, planning principles have also been decided on. First of all, it has been agreed that DVB-T will be the only standard to be dealt with in the conference as a representative for digital terrestrial television. Even though this decision reduces the planning effort significantly, it is well known that DVB-T allows for a huge number of different system variants. All of them require different planning parameters, which has a direct impact on the generation of a frequency plan.

This variety is a very important feature of DVB-T from which broadcasters are able to benefit when it comes to network planning. But on the other hand, it became clear during the preparations of the conferences that including every single detail of a system variant in a plan entry to a frequency plan constrains the freedom of network operators, in particular, in view of the freedom DVB-T offers as being a COFDM system allowing SFN operation. Hence, in order to simplify the planning process as much as possible three so-called reference planning configurations (RPC) have been developed (see Section 5.4). They act as some kind of placeholders representing a large number of different system variants requesting similar planning parameters. The three RPCs represent fixed reception, portable outdoor, or mobile reception and portable indoor reception. Furthermore, four different reference networks have been included, which allow to assess the interference a typical network is going to impose on other networks during the plan generation process. By choosing an appropriate combination of RPC and RN intended coverage targets could be mapped to mathematical objects that could be appropriately dealt with by the RRC-06.

Another important issue of the RRC-04 was to make appropriate provisions for T-DAB planning. CEPT had already made frequency plans for T-DAB in the years before (see sections above) and many T-DAB networks have been put into operation. From a CEPT perspective it was important to ensure that this could somehow be integrated into the new plan for VHF. Initial resistance against allowing T-DAB to make use of "television frequencies" could be overcome. Similar to DVB-T two RPCs and two RNs were proposed in the report submitted to RRC-06. One of the RPCs corresponds to the planning scenario of WI95, referring to mobile reception while the new RPC reflected the fact that also portable indoor reception would be important for the future development of T-DAB [Bru05].

The conference was characterized by sometimes heated discussion about fundamental concepts. Equitable access to the spectrum was one of them. Since there was no consensus, how this could be cast into quantitative terms, the discussion was postponed to the work between the RRC-04 and the RRC-06.

Furthermore, it became very quickly clear that in contrast to WI95 or MA02, not only allotment planning was requested by administrations. ST61 was an assignment plan and there were many administrations in the planning area, which wanted to base their input requirements to the new frequency plan on ST61. Basically, they were pursuing some kind of conversion strategy from analogue to digital broadcasting, which nevertheless, meant that they favored a planning approach based on assignments. Therefore, in the end it was decided that both types of requirements should be dealt with. This had direct impact on the preparations of the RRC-06 because a discussion started off as to which ERP requirements should be taken into account for planning. After lengthy and very controversial discussions, it was decided that stations with less than 50 W in Band III and less than 250 W in Bands IV/V should not be accepted for the planning process as input.

The idea to derive input requirements from existing plan entries of ST61 was relevant in relation to assignment planning. Also several administrations, which favored allotment planning wanted to exploit ST61. This gave rise to the development of what was called the channel

potential method. In particular, in bi- or multilateral coordination negotiations before the conference this proved to be a very powerful method
in order to derive a set of multilaterally agreed input requirements, which
were considered mutually compatible.

Basically, the idea was to identify the maximum extension of allotment areas inside a given country without the need for further bi- or
multilateral coordination. Figures 7.2 and 7.3 sketch the approach. In
Figure 7.2 several allotment areas are shown, which are assumed to be
located in the three neighboring countries A, B, and C.

The allotment areas are assumed to be mutually separated large
enough to use the same channel in all allotments. However, inside each
country the same channel could be used in other areas as well without getting in conflict with the usage in other countries. This is shown
in Figure 7.3 where the large dark areas enclosing the allotment areas
represent the so-called channel potentials of the considered channel.

Bands III, IV, and V are not exclusively used for broadcasting. There
are many other services, which are listed in the RR of ITU as primary
services. Therefore, administrations have to coordinate with affected
administrations whenever a broadcasting service or one of these other
services is newly introduced or an existing station is modified. Clearly,
the plan generation process for digital terrestrial broadcasting had to

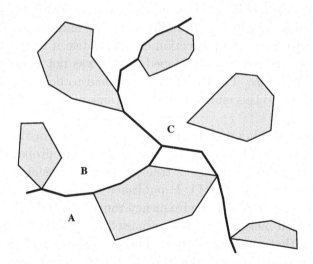

Figure 7.2: Allotment areas in three neighboring countries.

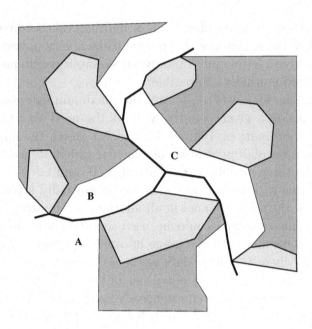

Figure 7.3: Channel potential areas derived from a set of allotment
areas in three neighboring countries.

take into account the protection of other services. Among them there
are fixed services, mobile services, aeronautical radionavigation services,
radio astronomy, broadcasting-satellite services, and others. During the
RCC-04, interference and protection criteria between digital broadcasting and these services were developed, but it was not possible to cover
all sharing situations. Therefore, this work had to be left to the intersessional period. To this end, several resolutions were issued by Committee 4 and 5.

Frequency planning rests on predictions of interference levels at given
geographical points. This calls for appropriate wave propagation methods to predict the field strength a transmitter produces at a given point.
Shortly before the RRC-04 ITU-R published a new recommendation for
field strength prediction in the frequency range 30–3000 MHz [ITU01a].
The propagation method that was proposed to be used for the RRC-06
was based on that recommendation. There were several deviations from
the original version, in particular in relation to the treatment of negative
effective transmitting antenna heights and the way in which mixed paths
propagations were dealt with.

The large planning area quite naturally brought together many very differently developed countries. In Europe, digital terrestrial broadcasting had already been introduced to some extend, while in Africa or the Arabic countries deployment of analogue television was still in full swing. Therefore, it is evident that there were different ideas about when and how the transition to an all-digital broadcasting world should be accomplished. During the so-called transition period, existing and planned analogue station would need to be protected. After that date, the protection rights of analogue services would cease. The date when the transition period should end, caused some conflict even during the preparation of the RRC-04. It was not possible to come to a common view on this issue, and hence, RRC-04 identified two options for the end of the transition period. In particular, CEPT-countries favored to cease analogue transmissions as soon as possible but no later than 2015. The other option said the end of the transition period should not be before 2028, but also not later than 2038. It was left to RRC-06 to finally decide on the end of the transition period.

The clash about the transition period triggered the development of a new element to be included in the digital terrestrial broadcasting plan. It is what was later called the envelope concept. At the beginning it was referred to as mask concept, which was later abandoned in order not to mix it with the idea of spectral masks for T-DAB or DVB-T emissions. Confronted with the fact that there was no consensus in sight, in relation to the transition period, the question was asked within CEPT if some measure could be developed, which would allow to continue the operation of analogue services even beyond the end of the transition period. The proposal was to operate the analogue station under the envelope of a digital plan entry. According to the planning principles, the technical characteristics of a plan entry precisely define the amount of interference that will be produced at any given point. Vice versa, it would be also clear how much interference this plan entry would have to accept at the boundaries of its own service area. The envelope concept simply stated that the operation of an analogue station would be feasible if not more protection is claimed nor more interference is produced as determined by the associated plan entry.

Finally, RRC-04 laid the foundations for the structure of the planning process itself. Basically, the same methodology as used in WI95 and MA02 was adopted, apart from necessary modifications due to the different systems that had to be taken into account. Therefore, the planning process consisted of the two steps compatibility analysis and plan synthesis, which could be iterated if necessary. More details on the RRC-04 can be found in [Pui04] and [Beu04c].

7.5   The Regional Radiocommunication Conference RRC-06 and the GE06 Agreement

The RRC-04 laid the foundations for RRC-06 by establishing the technical basis and providing the planning criteria and parameters for the new plan. Since during the RRC-04 not all tasks could be finished, the two years between the two conferences were a busy time for frequency planners. The intersessional period was used by ITU, CEPT, and EBU to fill those gaps that were left by RRC-04. At ITU level several working groups were established. This was foreseen by RRC-04 in terms of several Resolutions. Three of these groups were of particular importance, namely the Intersessional Planning Group (IPG), the Planning eXercise Team (PXT), and the Regulatory and Procedural Group (RPG). The IPG was to develop draft plans during the intersessional period, taking account of bi- and multilateral negotiations carried out by the administrations. PXT should test the planning software, which was provided by the technical department of the EBU by carrying out planning exercises. Finally, RPG had the task to prepare the regulatory and procedural framework, which was to be drawn up in the new agreement.

RRC-06 took place from May 15, 2006 until June 16, 2006. More than 1000 delegates from 104 different countries participated. As in the case of the RRC-04, the conference was structured by setting up six Committees dealing with different items. Again Committee 4 (Planning Committee) and 5 (Regulatory Committee) constituted the core of the planning conference. Committee 4 was subdivided into five so-called Coordination and Negotiation Groups (CNG), which were created to divide the entire planning area into smaller regions. They were defined in a way that these areas could be considered as more or less independent from each other. CNG1 covered Europe and the North Eastern part of the planning area. CNG2 covered Western and Central Africa, CNG3 covered Eastern and Southern Africa, CNG4 covered the Red Sea area where extreme propagation conditions are encountered, while CNG5 covered countries around the Mediterranean Sea. Committee 5 was dealing with regulatory, procedural, and technical aspects of using Bands III, IV, and V. In particular, the development of the agreement text was in the scope of Committee 5.

The technical department of the EBU had developed the software that was used for the planning process of the RRC-06. This covered both the compatibility analysis and the plan synthesis. The Radiocommunication Bureau of the ITU (BR) provided software tools to capture and validate the input data and to visualize the results of the planning process. A joined team of EBU and ITU experts carried out the intensive calculations during the conference. In particular, the compatibility analysis required so much computational power that the computing facilities of European Organization for Nuclear Research (CERN) [CER08] in Geneva had to be exploited.

The compatibility analysis was needed to identify whether two requirements for digital broadcasting could share a frequency or not. Furthermore, the impact of digital requirements on assignments to other primary services, including existing or planned analogue television assignments to be protected had to be evaluated as well. These calculations resulted in a list of available frequencies for each digital broadcasting requirement. The results of the compatibility analysis were taken into account during the plan synthesis, which was based on the implementation of several thousands of different graph theoretical algorithms for the frequency assignment process. They were run concurrently as much as possible by distributing the task onto the several hundreds of computers available at the CERN computer center.

7.5.1   Overview of the Results of GE06

A total of 118 administrations submitted requirements to be taken into account in the planning process. They were asking for frequencies, that is T-DAB blocks or TV channels, in the frequency ranges 174–230 MHz and 470–862 MHz. In the UHF Bands IV and V, a channel raster of 8 MHz was used throughout the entire planning area. However, in Band III the situation was more complicated. A mixture of different raster schemes needed to be taken into account. T-DAB employes spectrum blocks of 1.75 MHz bandwidth. This was unique, but for DVB-T both 7 MHz and 8 MHz had to be considered in Band III.

The first set of input requirements at the beginning of the conference comprised more than 80,000 data sets out of which roughly a fourth referred to VHF. This pile of input data was the starting point for the planning marathon. Planning was carried out iteratively in terms of four iterations. The overall objective of the planning activities was to satisfy as many requirements as possible, that is to find frequencies for them. However, it turned out that the number of submitted requirements in connection with the constraints imposed by the other primary services to be protected was far beyond what could be accommodated in the spectrum at hand. Therefore, administrations were having extensive bi- and multilateral coordination meetings in order to figure out under which conditions the incompatibilities between parts of their requirements could be overcome.

The coordination efforts resulted in a very large number of so-called "agreements by administrative declarations," which became a crucial means for the successful plan generation. Providing an administrative declaration basically meant that two or more administrations carried out detailed geographically limited studies on the basis of more developed and sophisticated planning tools than those agreed at the RRC-04 to be employed by RRC-06. Subsequently, requirements were declared mutually compatible thereby overwriting the formal findings of the compatibility analysis.

Very often it was clear for administrators and broadcasters that for digital terrestrial broadcasting networks the existing transmitters sites so far used for analogue broadcasting must be used for the implementation of the digital transmitter network due to economical reasons. In some cases, where this led to obvious interference conflicts between transmitter sites, administrators simply agreed to accept higher levels of mutual interference. Hence, they provided corresponding administrative declarations.

It has to be borne in mind that in the context of RRC-06, the level of interference was evaluated on the basis of the wave propagation method adopted, which was based on ITU-R Recommendation P.1546 [ITU01a]. However, special topographic and morphologic condition cannot be taken into account by that method. From their long-time experience of operating transmitters under very different conditions, broadcasters were quite aware of the mutual interference potential between different transmitters sites. For example, there are many cases in which terrain shielding basically leads to a decoupling of two transmitter sites so that they can share a channel in practice even though investigations based on ITU-R Recommendation P.1546 [ITU01a] might indicate that spectrum sharing is not possible. Furthermore, detailed calculations employing more developed and sophisticated prediction methods showed that spectrum sharing could be envisaged, if proper antenna design is considered.

In some regions of the planning area like parts of Africa or the Arabic countries, where no coordination had been carried out between administrations before the RRC-06 during the preparation phase, a general relaxation of the planning criteria by up to 5 dB was agreed in order to get requirements into the plan, which otherwise would have not gotten a frequency. In that cases, it was obvious that any implementation of a plan entry will need to be coordinated before bringing a station into operation.

One of the issues that lead to sometimes fierce arguments in the preparation of GE06 was the protection of analogue TV stations. RRC-06 was to draw up a frequency plan for digital terrestrial broadcasting in a spectrum range, which was more or less fully occupied. The planning activities carried out by the PXT during the intersessional period clearly showed that there is no hope to establish a new frequency plan, if the plan generation process would need to protect existing and planned analogue TV assignments. Consequently, RRC-06 quickly decided not to protect analogue stations during the plan generation process. This paved the way for a successful plan for DVB-T and T-DAB.

Even though analogue transmission would not be protected by the final digital plan, it was nevertheless clear that the introduction of digital terrestrial broadcasting would not be accomplished over night. A transition period during which analogue transmissions would be granted protection would be necessary. Basically, protection of analogue stations means that a digital station can only be put into operation after successful coordination with administrations whose analogue transmissions would be affected.

Opinions diverged heavily about the duration of the transition period. RRC-04 had already not been in a position to decide on the duration of the transition period. After very lengthy discussion, it was finally decided by RRC-06 that June 17, 2015 should be the end of the transitions period. As an exception from this rule, it was agreed that for some non-European countries 2020 should be applied as the end of the transition period for VHF. CEPT was of the opinion that 2015 would still be too far away and decided that 2012 should be valid for its members. Administrations were in addition encouraged to go digital as fast as possible. It should be noted that other primary services than analogue TV were fully protected for the generation of GE06. Therefore, the term transition period exclusively refers to the protection of analogue terrestrial broadcasting.

In order to know, which analogue stations would need to be protected during the transition period, it was necessary to establish a reference situation for analogue television. It was decided that for the territories governed by ST61 and GE89 the reference situation should be given by the corresponding updated frequency plans. Thus, all assignments successfully coordinated until March 15, 2006 would be included in the reference situation. For assignments to other primary services also a reference situation was defined, comprising all successfully coordinated assignments, which have been notified to the ITU at the same date. This decision had already been prepared by RRC-04 and therefore administrations knew since then that if they wanted to claim protection for analogue stations or assignments to other primary services they had to take the effort to bring their notifications to the ITU up to date.

The digital switch-over, how the transition from analogue to digital broadcasting is also called, was, and still is, a very complex enterprise. In many cases, neighboring countries have entirely different ideas about the time horizon of the transition as well as the manner in which it could be accomplished. However, this poses a problem since along national borders the transition strategies need to be aligned. Some administrations are in favor of a hard changeover, that is switching off analogue transmission at a defined date and starting digital transmissions seamlessly. Others want to introduce T-DAB and DVB-T by gradually switching from analogue to digital broadcasting on a transmitter by transmitter basis.

Those digital plan entries, which are not in conflict with any other co-channel users can be implemented in the short-term without problems. For those plan entries, whose operation is subject to the protection of analogue transmissions successful coordination between administrations has to be achieved prior to bringing digital plan entries into operation. However, it might, nevertheless, be possible to realize a partial network implementation. In the case of an assignment plan entry, this means to use reduced ERP or specially adapted antenna patterns for the time when analogue stations need to be protected. For allotment plan entries there is more freedom. An allotment can be implemented in terms of building only part of a full SFN in a first step. This means not the whole allotment area will be served but only that part which is far-off the existing analogue transmissions. This way it is possible to limit the interference to analogue transmissions to an acceptable level during the transition period.

The key to achieve consensus amongst the administrations participating at RRC-06, concerning the duration of the transition period was the development of the envelope concept. This idea had been put forward by RRC-04 already. Basically, it allows to make use of a plan entry for DVB-T or T-DAB for another system as long as not more protection is requested and not more interference is produced than the underlying digital plan entry would do. Clearly, "another system" could also be analogue television. A digital assignment of the GE06 plan could be used as a "placeholder" under which an analogue station can be operated even beyond the end of the transition period when the protection of analogue transmission would have ceased.

The envelope concept is described in Article 5.1.3 of the GE06 Agreement [ITU06]. It states that a digital plan entry might be used for systems, whose technical characteristics are different from those appearing in the plan. Both broadcasting services and also other primary services can be implemented. However, they need to be in conformity with the Radio Regulations of the ITU [ITU04]. This is in the first place a regulatory constraint, which means that only those systems can exploit GE06 under the envelope concept that are defined as primary services in the Bands III, IV, or V already. From a technical point of view, the basic condition to be met is that the peak-power density in any 4 kHz interval shall not exceed the spectral-power density in the same 4 kHz as produced by the digital plan entry. Even though originally intended as a door opener for those administrations, who were not in favor of 2015 as the end of the transition period, it turned out that the envelope concept developed some momentum in a direction not well received by broadcasters. This aspect will be further discussed in Chapter 8.

The generation of the GE06 frequency plan was accomplished after four planning iterations. Administrations had to submit their requirements at a given time, usually, before the weekend. They were given a bit more time to prepare the administrative declarations, which were taken into account only after the full pairwise compatibility analysis had been carried out. This information was taken into account during the plan synthesis process. The large number of inputs to the first iteration resulted in only approximately 65% and 74% of satisfied requirements in VHF and UHF, respectively.

From one iteration to the next, administrations were urged to keep the number of changes of their input data to a minimum, and for the fourth iteration only corrections of mistakes to the already existing input data sets like typos were accepted. The only modifications to the input that were always accepted was a reduction of the number of input requirements and the provision of more administrative declarations in order to make channel sharing possible. In the end, the total number of requirements for digital terrestrial broadcasting had dropped from roughly 80,000 to about 72,000 and the number of satisfied requirements increased to 93% and 98% in VHF and UHF, respectively.

The RRC-06 planning process provided three sets of results, namely a digital plan for T-DAB and DVB-T containing both allotments and assignments, a frequency plan for analogue television that need to be protected during the transition period, and a list of assignments to other primary services in the frequency bands under consideration. In particular, in CEPT countries allotment planning was favored. In the end, the majority of CEPT countries obtained seven national coverages of DVB-T in UHF and three or four layers for T-DAB and DVB-T in VHF.

The majority of DVB-T plan entries, roughly 65%, refers to fixed reception, which in the case of allotments is represented by RPC1. The rest corresponds almost entirely to portable outdoor reception (RPC2) because the number of portable indoor reception entries (RPC3) is almost negligible compared to the other two cases. For T-DAB, the situation is similar concerning the two possibilities, namely mobile (RPC4) and portable reception (RPC5). Also, here the number of mobile allocations is almost twice as large as that of portable indoor reception. Concerning the statistics of allocated frequencies, it has to be noted that lower channels in UHF have been requested more often and subsequently allocated more frequently. In particular, starting with channel 61 the channel usage significantly decreases. This can very likely be addressed to the fact that in many countries of the GE06 planning area this spectrum range is allocated to other primary services as well. Therefore, this part of the spectrum was not available for digital terrestrial broadcasting. In the case of Band III, DVB-T allocations are primarily in channels 5–10 for the 7 MHz raster. This is due to the fact the WI95 allocations for T-DAB, which most administrations wanted to be included in the new plan as well, were concentrated in channels 11 and 12.

As discussed above, administration provided a vast number of administrative declarations in order to get their requirements in the plan.

As a consequence, implementation of these plan entries will be subject to the conditions agreed by the concerned administrations. Basically, there are three different types of conditions to be taken into account. For some entries coordination with respect to existing or planned analogue TV stations is explicitly required, before the digital plan entry can be brought into operation. The second category refers to special conditions agreed by administrations in relation to other digital broadcasting plan entries. Actually, this corresponds to the standard case of getting broadcasting requirements into the plan, which failed the compatibility analysis. Finally, there are cases, where conditions have to be met relating to the protection of other primary services, where administrations had agreed on special measures to be taken. In any case, the details of the agreement between administrations are not contained within the GE06 plan. It includes only remarks that there are some constraints to be considered. Administrations take the responsibility themselves to be able to retrieve the proper wording of their agreements. A more detailed analysis of the results of RRC-06 in terms of channel usage and geographical distribution of channels can be found in [OLe06]. Clearly, the most comprehensive source of information is the ITU-R Website dealing with GE06 [ITU07a].

7.5.2   Article 4 — Plan Modification Procedure

Setting up the GE06 plan was certainly a big task. However, it was also clear that the new plan will not be static. There will be changes to it, either new plan entries need to be added, existing ones might be modified, or some might even be deleted. In any case, it was one of the objectives of RRC-06 to provide the regulatory and technical means for all these intentions. Committee 5 took care of that very difficult and politically delicate task. To this end, Article 4 of GE04 was prepared to contain procedures, which specify in detail what has to be done by an administration wishing to make a change to the GE06 plan. Any addition or modification will very likely have an impact on other digital-plan entries, assignments to other primary services, and during the transition period will need to protect existing and planned analogue TV assignments. Consequently, coordination with all affected administrations need to be sought.

Originally it was intended to base the new frequency plan on two planning objects, namely assignments and allotments. Already this decision had been an innovation, since in the past either assignment plans have been drawn up (ST61 and GE89) or allotment plans were established (WI95 and MA02). Therefore, mixing assignment and allotment planning was already a step into uncharted waters. However, during the preparation of the RRC-06, it turned out that administrations had very different demands that could only be satisfied by introducing a total of five different planning objects identified by five distinct plan entry codes.

The simplest planning object is a single assignment. This reflects exactly the same way of planning as in ST61 or GE89, that is an administration provides all technical characteristics of a transmitter site, which are necessary to assess its interference impact at any given geographical location. A natural extension of a single assignment is a set of assignments that are combined by the same SFN identifier. The intension is to use them in SFN mode, that is using the same frequency to broadcast the same content. Then, a single allotment could be provided as input to the planning process. It is given in terms of a set of geographical vertices, defining an area to be served by an appropriate transmitter network. An allotment is associated with a RPC and a RN. This information is sufficient to assess the interference produced by the allotment and the protection it needs to be granted. Plan entry code 4 refers to an allotment with linked assignment(s) and a SFN-ID. Linking assignments was invented to bring together the concept of allotment with existing transmitter sites. Employing such a construct allows to take into account the features of transmitters, which are intended to implement the allotment but can however, not be represented by the characteristics of the allotment, that is RPC and RN. Finally, a fifth planning object was introduced, which consists of an allotment to which a single assignment is linked, but no SFN-ID is provided. The intention of this was to combine the concept of a defined service area as given by an allotment area and special transmission characteristics that can be modeled by providing a corresponding assignment.

Since the five different plan entry codes have very different properties, it is evident that they need to be treated differently. Both Article 4 and 5 procedures required to calculate field strength values produced by these planning objects at given points. To this end, it has to be clearly defined what the source of interference for each of these cases is. For a single assignment and a single allotment the situation is obvious. The technical characteristics of the assignment and the RPC/RN combination will be used, respectively. However, for the mixed cases field strength values based on both information has to be calculated for a given geographical point. Then, the larger value is defined as the field strength produced there by the planning object.

Another important aspect for both Article 4 and 5 is the definition of a point of reference for each plan entry code. Both GE06 Articles 4 and 5 employ geometrical concepts on which their calculation methods are based. Depending on the task, a set of geometrical contours has to be calculated. To this end, a geographical point of reference is needed. In the case of a single assignment, it is straightforward to use the location of the transmitter site thereto. However, already for an allotment this becomes a tricky issue, in particular, because a single allotment can consist of several distinct polygons. According to the GE06 rules this can be employed in order to more accurately treat main land and off-shore islands as a single allotment. In principle, the center of gravity of the allotment is used as the point of reference. In case of a set of assignments more sophisticated rules apply. More details can be found in [EBU07].

The objective of the procedure anchored in Article 4 is to determine which services could be potentially affected by a proposed modification or addition to the GE06 plan. Clearly, this means to identify the administration to which a particular service is associated. Services refer to both broadcasting services and other primary services. In order to limit the computational effort to a reasonable amount, first of all a contour separated by 1000 km from the location of the proposed addition or modification to the GE06 plan is constructed. Administrations whose territory falls into the identified area will have to be considered during the Article 4 procedure.

Any modification to the GE06 plan has to refer to one of the five plan entry codes. Depending on that the corresponding point of reference is determined. In case of assignment(s), the distance is measured from the transmitter site(s), while for an allotment the 1000 km are taken from the allotment boundary. Any country, whose border is intersected is taken into consideration. Furthermore, the frequency for the new or modified plan entry is important for the subsequent analysis. Figure 7.4 sketches the situation. It shows a broadcasting allotment to be added in country A. The 1000 km contour—from which only a small part is indicated here on top of the figure — intersects with countries B, C, and D, so they will have to be considered as potential candidates with whom coordination

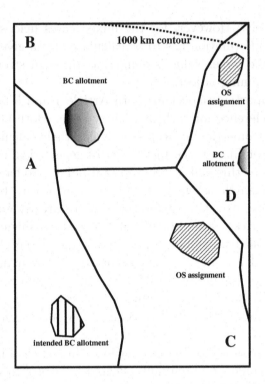

Figure 7.4: Thousand kilometer contour derived from a broadcasting allotment to be added to the GE06 plan according
to an Article 4 procedure.

might be required. In countries B,C, and D, there are broadcasting plan
entries and assignments to other primary services that are indicated in
terms of the corresponding service areas.

Basically, a modification of GE06 can have an impact on broadcasting services and/or other primary services. The identification of
administrations with whom agreement might need to be requested is
slightly different for these services. However, in any case so-called coordination contours are calculated that constitute the measure to decide, if
an administration needs to be approached for agreement or not. In principle, a coordination contour represents a curve along which the proposed
plan modification will create a certain field strength level, namely the
so-called coordination trigger value. The actual value of the trigger field
strength depends on the technical characteristics of the intended plan
modification and the type of service potentially affected. Therefore, the
identification of affected administrations has to be carried out in relation

to broadcasting plan entries in GE06 and against assignments to other primary services contained in the list attached to the GE06 Agreement. So, this might require to construct several different coordination contours.

In any case, there will be a contour relating to broadcasting. Different trigger field strength values for T-DAB and DVB-T are given in GE06. Till the end of the transition period, analogue television will need to be considered as well. The trigger values differ according to the frequency band, that is III, IV, or V. In order to simplify the calculations, only the most critical, that is smallest trigger value is employed to construct a single broadcasting coordination contour. If this contour intersects or encloses the national boundaries of a country, which is located within the 1000 km contour coordination with the corresponding administration is required.

In the case of other primary services, the details of the identification process are different. In a first step, those other services are identified whose assignments are located within the 1000 km contour. From these, only those will be taken into consideration that are contained in the list attached to the GE06 Agreement. Then, it is checked whether there is a frequency overlap between the intended plan modification and the frequency used by the assignments to other primary services. Only if there is an overlap, these systems will be taken into account. It is important to note that this constitutes a significant difference between the treatment of broadcasting and other primary services. For the latter, basically only co-channel usage can trigger coordination between administrations.

Once the assignments to other primary services have been identified the relevant trigger field strength values are extracted from the GE06 Agreement and the corresponding coordination contours is generated. In contrast to the broadcasting case not the national boundaries are relevant, but coordination with an administration is required if the locations of the receiving stations or the service areas of these other primary services are intersected or enclosed by the coordination contour. For further details it is referred to [EBU07].

Article 4 of GE06 contains the regulatory framework according to which an administrations has to act when wishing to modify the GE06 plan. All technical issues in relation to Article 4 are given in Section I of Annex 4 to the GE06 Agreement. It specifies in detail how the coordination contours are constructed for a given assignment or allotment. Apart from the explanation given in the GE06 Agreement, a very good description of the relevant technical issues seen from a practical point of view can be found in [EBU07]. Figure 7.5 illustrates the different coordination contours.

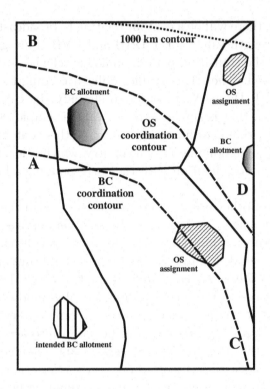

Figure 7.5: Two coordination contours derived from a broadcasting allotment that has to be added to the GE06 plan
according to an Article 4 procedure.

It is assumed in Figure 7.5 that there is only one type of other primary service inside the 1000 km contour. In that case, two coordination contours have to be constructed, one for broadcasting and one for the other services. The broadcasting contour intersects with countries B and C. Therefore, coordination concerning broadcasting services is required with these two countries. In the case of other primary services, the service area or the location of the transmitter site is relevant, not the national border. This means that coordination is required with country B in relation to other primary services and not with country D.

7.5.3 Article 5 – Notification Procedure

Setting up a frequency plan and having the means to modify existing plan entries gives administrations and broadcasters the possibility to operate transmitter networks for broadcasting services. However, it is necessary to introduce rules which guarantee that the implementation of a network is carried out in conformity with the characteristics of the plan entries. This means that the network implementation must not produce more interference than the associated plan entry will do. To this end, Article 5 has been included in the GE06 Agreement, which contains the technical details how the conformity check between plan entry and network implementation has to be carried out.

The operation of a transmitter under the GE06 Agreement calls for two requirements to be fulfilled, namely there must be a plan entry in GE06 with which the operational transmitter can be associated, and furthermore, an assignment corresponding to the technical characteristics of the transmitter has to be recorded in the ITU Master International Frequency Register (MIFR). The latter process is called "notification."

Basically, there are three cases in which an Article 5 procedure is needed. Two of them refer to particular modifications of plan entries resulting from an Article 4 procedure and one is a "true" Article 5 issue. The latter refers to the straightforward intention of an administration to use a particular plan entry in order to implement a transmitter network. An implementation can be accomplished in terms of one or several assignments. The technical characteristics of the assignment(s) are fed into the machinery of the conformity check in order to prove that the intended network will not produce more interference than is calculated on the basis of the plan entry characteristics.

Furthermore, Article 5 has to be employed in relation to an allowed Article 4 modification of an allotment plan entry, namely the conversion of an allotment plan entry into a set of assignments. This idea was taken over from the WI95 [CEP95] and MA02 [CEP02] Arrangements of CEPT relating to T-DAB. In principle, this means to substitute the original allotment plan entry by a number of assignments that then will become part of the plan. Such a plan modification is only allowed if the aggregated interference of the set of transmitters does not exceed the limits imposed by the allotment plan entry. This is assessed by applying the conformity check defined by Article 5 of GE06.

The second Article 4 case where reference to the conformity check is made concerns a plan modification, which claims to produce less interference than the original plan entry it refers to. Also in such a situation a check is needed in order to confirm this claim.

As in the case of Article 4 of GE06 concerning an intended plan modification, the regulatory framework an administration has to apply when submitting a notification to the ITU is given in the main body of the GE06 Agreement under Article 5, while all technical issues are presented in Section II of Annex 4 to the Agreement.

Application of an Article 5 procedure starts with the submission of a set of assignments, whose technical details have to be specified. Then, the conformity check of Article 5 comprises two examinations. First of all, the frequency and the location of the submitted assignments are checked. Clearly, the frequency has to be the same as that of the plan entry. Moreover, the location of the transmitter sites have to be close to the location of the plan entry. For an allotment this means that they are allowed to lie inside or outside the allotment area. The latter case is only acceptable, if the transmitter sites are separated from the allotment boundary by not more than 20 km. In the case of an assignment, the transmitter location might deviate by 20 km from the geographical location recorded in the plan.

The second examination of the conformity check relates to the technical characteristics of a plan entry, which allow to calculate a corresponding interfering field strength value at an arbitrary geographical point. The field strength values at all possible points define an interference envelope of the plan entry. An implementation is considered as being in conformity with GE06, if the network implementation stays below that interference envelope.

As already mentioned in the previous section, the GE06 comprises five different types of plan entries distinguished by their plan entry codes. Thus, it comes as no surprise that all of them require a slightly different treatment for this comparison. For any type of plan entry, it is possible to calculate a field strength value at any arbitrary geographical point. Therefore, the concept of an interference envelope is naturally a two-dimensional concept. In other words, the interference envelope of a plan entry can be imagined as a two-dimensional surface above a given area. This is just like a mathematical function of two independent variable x and y corresponding to geographical latitude and longitude. The height of the surface above any point represented by a pair of longitude and latitude values is then given by the field strength value produced there by the plan entry. Checking if a network implementation is in conformity with a plan entry would consequently require to check any point throughout a given area. Since this is not feasible from a practical point of view, a set of calculation points is defined where a comparison of the field strength produced by the plan entry and by the intended network implementation is carried out.

In order to limit the amount of computational effort, the area in which calculation points are located is bounded by a so-called cut-off contour. The cut-off contour is basically a trigger field strength contour similar to those used in the application of the Article 4 procedure. However, in contrast to these, there is only one cut-off contour that takes into account both broadcasting and other primary services appropriately. The first step of the construction of the cut-off contour is the determination of the point of reference of the plan entry for which a particular implementation is to be assessed. For a single assignment this point corresponds to the location of the assignment. In case a set of assignments is dealt with, the center of gravity of all locations is chosen while as soon as an allotment is involved the center of gravity of the allotment area is used. Then, radials are drawn every 1°, starting at the point of reference and extending to infinity. Along these radials the point is identified, where the field strength produced by the actual implementation of the plan entry reaches the broadcasting trigger field strength as defined in Article 4. Just to avoid any misunderstanding, the cut-off contour is calculated on the basis of the technical characteristics of the transmitter(s) submitted for notification. In the case of an allotment, already notified assignments that have been entered into the MIFR already have to be included, since they make a part of the network implementation of the allotment. Clearly, this is also valid for the comparison in each calculation later in the process.

All points found that way are subsequently connected and the resulting polygon defines the cut-off contour. However, there are additional conditions to be taken into account. If the constructed contour lies entirely inside the territory of the administration whose plan entry is considered, then other primary services have to be accounted for as well. This might lead to a modified cut-off contour, whose calculation is then based on the trigger values of other primary services. Annex 2 of [EBU07] explains in details all possible implications of different conditions on the construction. Figure 7.6 sketches a simple layout for the case of a single assignment and a single allotment plan entry.

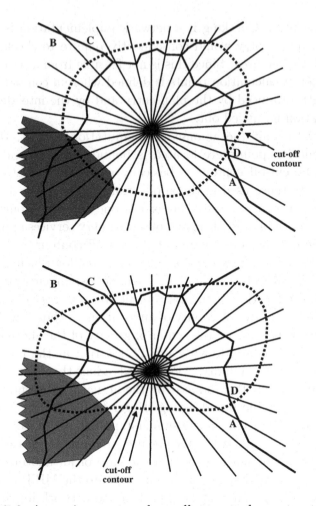

Figure 7.6: An assignment and an allotment plan entry together
with the corresponding cut-off contours needed for the
application of an Article 5 procedure. The cut-off contours are meant as examples. They depend on the
technical characteristics of the plan entries, that is
RPC/RN or ERP, antenna height and antenna diagram, respectively. The dark areas on the lower-left
represent sea.

Once the cut-off contour has been generated, the calculation points
have to be identified. As a general rule, calculation points must lie inside
the area delimited by the cut-off contour. At the same time only those

calculation points are considered that are located outside the territory of the administration whose Article 4 or 5 request initiates the conformity check. This means that there can be cases, where no calculation points are found at all. In that case, conformity with the GE06 plan is granted by definition.

If the cut-off contour extends beyond the national boundary, the location of the calculation points have to be determined. To this end, a set of geometrical contours is generated. They constitute contours that are separated from the location of the plan entry by a constant distance, in particular the distances 60 km, 100 km , 200 km, 300 km, 500 km, 750 km, and 1000 km are employed. For assignment plan entries, the contours are concentric circles around the geographical location of the assignment, which at the same time acts as point of reference. In the case of an allotment, the geometrical contours correspond to buffer zones around the allotment area. Figure 7.7 shows some geometrical contours.

The point where the radials emerging from the point of reference of the plan entry under consideration and the geometrical contours intersect are the locations of potential calculation points. The word "potential" is to indicate that not all intersection point are employed. Rather, only those points lying outside the national territory of the notifying administration and inside the cut-off contour are utilized as calculation points. Whether they are located on land or above water is not important, both are taken into account. Due to the fact that the radials are separated by $1°$ and that there are seven geometrical contours the maximum number of calculation points that ever might need to be considered is $360 \times 7 = 2520$. However, it can be expected that in practice the number will be significantly less. Figure 7.8 presents the locations of the calculation points for the examples considered here.

The technical implementation of Article 5 as given in Section II of Annex 4 of GE06 exhibits one very important feature. The location of the calculation points is derived exclusively from characteristics of the plan entry. Both point of reference and circular contours, or buffer zones are calculated from the geometrical features of the assignments or allotments in the plan, respectively. However, the cut-off contour is determined on the basis of the technical characteristics of the intended network implementation by calculating field strength levels and comparing them with trigger field strength values. This is reasonable from the point of view that, for example, a high-power assignment plan entry could be implemented in terms of low-power station. This will certainly

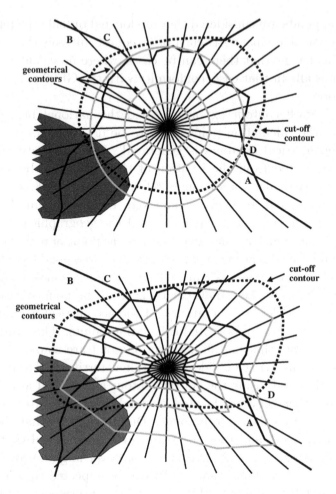

Figure 7.7: An assignment and an allotment plan entry together
           with the corresponding cut-off contours and geomet-
           rical contours. The dark areas on the lower-left repre-
           sent sea.

have a rather limited impact on other plan entries or already existing
networks. Consequently, it is not relevant to carry out intensive calculations at distances, where the field strength values produced by the
intended network implementation has fallen already to irrelevant levels.

Once the location of the calculation points has been determined the
actual assessment, if the network implementation is in conformity with
the characteristics of the plan entry can be accomplished. To this end,

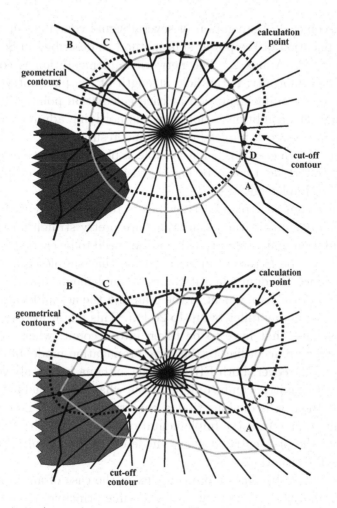

Figure 7.8: An assignment and an allotment plan entry together
           with the corresponding cut-off contours, geometrical
           contours, and calculation points. The dark areas on
           the lower-left represent sea.

two calculations are carried for each calculation point. First, the field
strength value is calculated that will be produced by the plan entry at the
given calculation point. Then, the technical characteristics of the assign-
ment(s) representing the intended network are used to evaluate the field
strength, which is produced at the same point. All calculations are based
on the wave propagation model described in Chapter 2 to Annex 2 of the

GE06 Agreement. Aggregation of several signal contributions is based on the application of the power sum method as described in Section 3.5 of Chapter 3 to Annex 2. The network implementation is considered as being in conformity, if the field strength of the plan entry is larger than that of the implementation at each calculation point. If only at a single point this condition is not met, the notifying administration has to modify its network implementation and resubmit the new technical characteristics. In case this is not feasible, a plan modification procedure according to Article 4 could be envisaged in order to match plan entry and network implementation.

For an assignment plan entry, which is to be implemented in terms of a single assignment, the calculations are pretty straightforward. For each calculation point only two calculations need to be carried out. However, it should be borne in mind that even that simple case allows for relatively much freedom when it comes to implementation. First of all, the location of the assignment in the plan and the actual location of the transmitter site may differ by up to 20 km. This might lead to different effective heights that have to be taken into account, when calculating the field strength according to the wave propagation model of the GE06 Agreement. Moreover, both ERP and the antenna height above ground of the real transmitter can be different from what is recorded in the GE06 plan. But then, it is obvious that a conformity check is mandatory in order to guarantee that the implementation stays below the interference envelope of the plan entry at every single calculation point. Figure 7.9 illustrates the situation.

The situation becomes more complex in the case of an implementation of an allotment plan entry. Still the same principle applies, namely to compare the interference produced by the plan entry with that of the intended network implementation. However, due to the larger freedom allotment planning offers, the computations are more elaborate. An allotment is represented by a combination of RPC and RN (see Sections 5.4 and 5.3). They define the interference envelope of an allotment. At any arbitrary point, a field strength value can be calculated by properly adjusting the RN along the allotment boundary. It is positioned at each vertex of the allotment polygon[11] and the aggregated field strength produced by the set of transmitters in the RN is computed at the point of reception under consideration. Then, the maximum value obtained is

---

[11]The details about the positioning of the RN at a given vertex can be found in [ITU06] in Appendix 2 to Section II of Annex 4 to the GE06 Agreement.

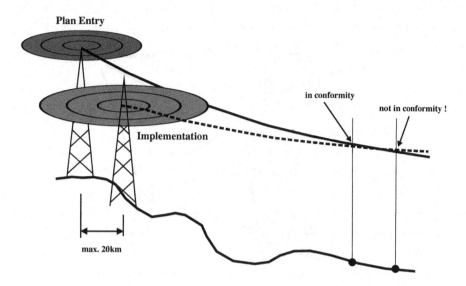

Figure 7.9: Characteristics of a plan entry and an intended implementation of an assignment can differ significantly,
which calls for an appropriate conformity check even
for this simple situation.

defined as the field strength the allotment produces there. These calculations are repeated for each vertex. Figure 7.10 visualizes the layout.

It has to be noted that the orientation of the RN relative to allotment
boundary depends on the direction of the line connecting the calculation
point and the vertices of the allotment. As a consequence, there are
situations, where the transmitters of the RN will lie outside the allotment
area. They might be located in the territory of an adjacent country or
in the sea. This caused a lot of heated discussion during the RRC-
06. In the end, administrations could agree to that concept noting that
the same mechanism is used in the plan-generation process during the
compatibility analysis, and therefore, plan generation and conformity
check are consistent.

Allotment planning gives the freedom to use as many transmitters
for the network implementation as considered appropriate by an administration. From a principle point of view, there is no limit as long as
the total interference of such a network stays below the interference envelope of the corresponding allotment plan entry. Nevertheless, at the
stage of implementation the technical characteristics of the transmitters
need to be specified so that they can be used for the conformity check

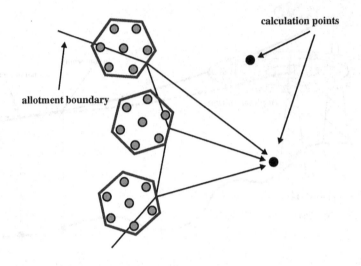

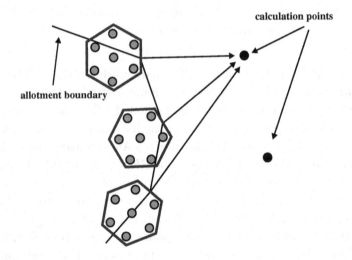

Figure 7.10: Calculation of the interference of an allotment plan
entry by positioning a RN along the vertices of an allotment boundary. The orientation of the RN varies,
depending on the direction of the connecting line between calculation point and vertex.

under Article 5. The set of notified transmitters is employed to calculate the aggregate field strength produced at the same calculation points as before. Figure 7.11 shows the scenery corresponding to the computation of Figure 7.10.

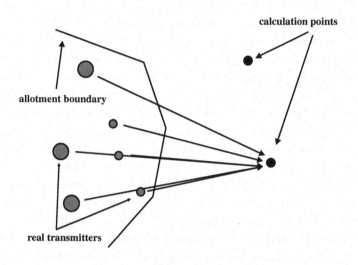

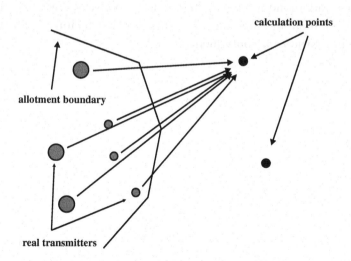

Figure 7.11: Calculation of the interference produced at given calculation points by transmitters implementing an allotment plan entry.

Already during the preparation of the RRC-06 and at the RRC-06 there were heated discussion about the conformity check. It was considered as too complicated and computationally challenging. Even if this might be true, it has to be noted that the decision to base the plan generation on assignments and allotments enforced the development of a methodology, which is flexible to cope with all potential implementation situations. The idea of an interference envelope under which administrations can virtually do whatever they like seemed to be very attractive in view of the rapidly changing digital telecommunications and broadcasting sector, leaving enough freedom to adapt to future demands. In the first place, this required to replace the one-dimensional analysis along a given curve or polygon as included, for example, in the WI95 Arrangement of CEPT by a two-dimensional approach.

However, the issue became complicated when some administrations demanded the introduction of a cut-off contour. This led to plenty of conceptional problems. Still it has not been fully proven that the way the cut-off contour is now calculated does indeed properly cover any eventuality. The original proposal not to use a cut-off contour, that is not to reduce the number of calculation points by suppressing some of them might have established a more sound and transparent basis for the conformity check. However, any measure included in the GE06 Agreement has undergone a lengthy and tedious political consensus building process. So, it should not come as a surprise that technical shortcomings have to be borne as a consequence.

Chapter 8

Future Developments

The establishment of the GE06 Agreement was a process, which took up several years, keeping hundreds of frequency planners from administrations and broadcasting companies very busy. In the end, a new frequency plan had been drawn up for digital terrestrial broadcasting, which was represented by two particular systems, namely T-DAB and DVB-T. Both systems allow for a large number of different operation modes targeting at very different reception modes and available data capacity. By developing the concept of RPCs and RNs this great variety was cut down to a manageable number of "placeholders", which were used during the plan generation process.

It was obvious to everybody that while the broadcasting community was preparing the RRC-06, the technical development was not put on halt in the meantime. Several new broadcasting systems or offsprings of existing ones have been presented over the years. Chapter 2 gives an overview about the most important ones to date. However, this list is clearly not exhaustive. Even if not being of primary importance at the time of the RRC-06, the question whether the new plan would leave enough freedom to adapt to future developments, be it more demand for broadcasting or for new systems, was always somehow present.

8.1 The Envelope Concept

One of the real hot topics of the RRC-06 was the transition period throughout which analogue television stations would still be protected. Already before the preparatory conference RRC-04 there seemed to exist as an insurmountable dissent between the CEPT countries and countries

R. Beutler, *Digital Terrestrial Broadcasting Networks*,
DOI 10.1007/978-0-387-09635-3_8, © Springer Science+Business Media, LLC 2008

from Africa and the Arabic world. CEPT wanted to go digital as soon as possible while many of the other countries were still in the stage of deploying analogue transmitter networks. Consequently, their time horizon was totally different.

Confronted with this clash of interests, planning experts within CEPT started to think about a possible way to resolve the situation. The discussions culminated in the attempt to answer the question: "Under what conditions would it be possible to operate an analogue station even beyond the end of the transition period without the need to claim for special protection or ask for particular conditions?" The solution to this was the development of what was later called the envelope concept.

The idea is pretty simple and when looking at the planning principles adopted in terms of allotment planning, it is also very straightforward. First of all, using a plan entry under which analogue television could be operated implies that this can only be possible if such an operation does not claim more protection nor produces more interference than the plan entry it refers to. This stems from the fact that the application of Article 5 in order to assess a network implementation concerning its plan conformity starts from the assumption that a plan entry represents a certain interference envelope. This means that based on the technical characteristics of the plan entry it is possible to calculate a field strength value that will be produced at a given geographical point. As explained in Chapter 7, a network implementation is in conformity, if it stays below that interference envelope at all points considered. In a similar way, an analogue station could be packed under the interference envelope of a digital plan entry. To this end, both assignment and allotment plan entries could be employed. The idea is very intriguing, since the transition period problem would be solved instantly. The conference could decide on an arbitrary date for the end of the transition period and those administrations wishing to continue the operation of their analogue stations could simply carry on. However, as always a closer look revealed that the devil is in the details.

RRC-06 decided that analogue television stations should not be protected during the plan-generation process. This means that at the same geographical location where there is an analogue assignment the same channel could be given to an assignment or allotment requirement for digital broadcasting by the planning process without formally causing an incompatibility. This was possible because the analogue assignment did not exist for the planning process. However, the digital plan entry cannot be implemented until the end of the transition period or until the analogue station is switched off before that date. If there is no geographical overlap between the service areas of analogue and digital assignments or allotments, then it might be possible to bring a digital network in operation but only while protecting the analogue transmission. As a consequence, the achievable coverage can be rather poor compared to what would be possible in the all-digital scenario.

An administration, intending to continue to operate analogue TV beyond the end of the transition period would need to make sure that a digital plan entry exists, which could be used to mask the analogue transmission. The determination of an appropriate digital placeholder for the analogue station turned out to be nontrivial. The planning process only took into account pairwise interaction between co-channel users. In a real environment, however, there are always several co-channel interferers that have an impact. This means that the size and shape of the area served by the analogue station is determined by the combined interference of all relevant co-channel interferers. If a digital assignment is employed to mask the analogue transmission, then the digital service area results from the interference of all digital interferers. This must not necessarily match the existing analogue coverage. Even the idea to approximate the analogue service area by an allotment area, which by definition is to be protected does not guarantee that the analogue station will be able to serve this area in the presence of digital interferers.

Another problem popped up immediately in the discussion. In the context of Article 5, the comparison of the interference level between the plan entry and the network implementation at a given point refers to the same system, that is either T-DAB or DVB-T. So, the application of an accumulated quantity, such as the field strength in order to assess the situation is possible. Basically, the field strength is proportional to the integral over the spectral density of the signal. Therefore, different spectral densities associated with different systems might lead to the same field strength levels. However, the actual interference might be totally different. The fact that the energy in the interfering signal is concentrated in a rather small part of a TV channel will have a more severe impact on the wanted signal in that particular spectral range while outside that interval the impact will be less. In total, the interference will not be same as for a flat spectrum. Figure 8.1 shows an example of the two different spectral densities, the light one could represent a DVB-T signal, while the dark density resembles an analogue TV signal.

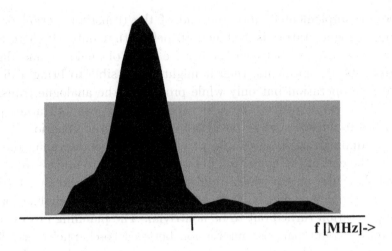

Figure 8.1: Schematical representation of two different spectral densities giving rise to the same field strength level.

In principle, there are two possibilities to define the conditions under which the envelope concept can be applied. The first solution is to say that the analogue signal has to be reduced in power to such an extent that its spectral density fits entirely under that of the DVB-T plan entry. Figure 8.2 sketches that idea. This would assure that the interference produced would be less than what is allowed by the DVB-T plan entry.

The other possibility is to exploit the characteristics of analogue TV explicitly in order to define what field strength levels would be allowed at the calculation points if the DVB-T plan entry would be implemented

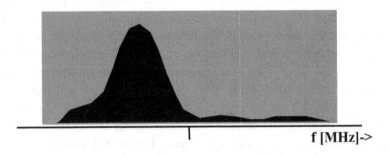

Figure 8.2: Spectral density of an analogue television signal squeezed under that of a DVB-T plan entry.

in terms of analogue TV. Basically, this would mean to make use of the protection ratios between DVB-T and analogue television.

Both approaches have advantages and disadvantages. Demanding to squeeze the analogue signal under the spectral density of DVB-T is simple and straightforward. However, it would probably limit the application of the envelope concept for the operation of analogue TV significantly. When comparing a 100 kW ERP DVB-T signal with a 100 kW ERP analogue signal, one can see that the spectral amplitudes differ roughly by 9–10 dB. Therefore, the ERP of the analogue transmission would need to be reduced by roughly 9–10 dB in order to fit the spectrum underneath the DVB-T spectrum. This would lead to a significant loss of analogue coverage. However, the other option for the application of the envelope concept has certain drawbacks. Using protection ratios is a clearly defined but very specific approach. It can be applied on a case-by-case only basis. On the other hand, the coverage loss expected could be more limited.

Once, there was an idea in the world that a DVB-T plan entry can be used for analogue television, it was natural to think about other possible usages of GE06 plan entries. At the RRC-06, DVB-H was not considered, nevertheless, it was already an important topic. Therefore, this was put forward as a potential other system to be put into operation by means of an application of the envelope concept. But this was just the start of the discussion. Since Band III was planned, both for T-DAB and DVB-T, it was natural to address the issue of implementing T-DAB under DVB-T. In Band III a TV channel comprises four T-DAB blocks and consequently a single DVB-T plan entry can be used to implement four T-DAB networks. Even the inverse approach was discussed, namely to use four contiguous T-DAB blocks in order to operate a DVB-T network under their joint envelope. At the time of the RRC-06, this idea was not further pursued, however, the revision of the MA02 Arrangement in 2007 [CEP07a] provided the means for such a usage of T-DAB blocks for the first time. Finally, it became clear that, in principle, any other terrestrial service occupying a bandwidth that does not exceed the 7 or 8 MHz of a DVB-T signal could be operated under the envelope of a GE06 plan entry. This was something that was not very welcome by broadcasters, because it was understood as a door opener for other services to a spectrum range, which so far had been exclusively used by broadcasting services.

In view of such a great variety of different usages of digital plan entries, RRC-06 decided that the rather strict approach demanding to fit the other service under the spectral density of the plan entry would be the only feasible one. Therefore, Paragraph §5.1.3 was included in the GE06 Agreement, which specified the conditions under which the envelope concept can be applied. It states that a digital plan entry "may also be notified for transmission in other primary services operating in conformity with the Radio Regulations if the spectral density averaged over 4 kHz interval across the entire bandwidth of the TV channel or T-DAB block does not exceed the spectral density in the same 4 kHz intervals of the plan entry" [ITU06]. This implies that there is no increase in interference to any other entry in the plan, nor to any other assignment taken into account in the preparation of the plan. Furthermore, no protection of the implemented system is sought, which would impose additional restrictions on any entry in the plan at the time the alternative system is implemented.

One particular part of the formulation of Paragraph §5.1.3 caused highly controversial discussions during the RRC-06, namely that the operation of other primary services has to be governed by the RR. Literally this means that only those other services can make use of the envelope concept, which according to the Frequency Table of Allocations of the RR are listed as primary services. In particular, CEPT administrations were of the opinion that this is not necessary. However, seen from an ITU perspective this is crucial. Any usage of a plan entry is subject to notification. An assignment can only be notified, if there exists a corresponding allocation of the service under consideration. Without that an administration could only notify broadcasting assignments under the plan entry that is to be used for another service. But then the question arises, how it is possible to assess whether this usage indeed obeys the constraints imposed by the envelope concept.

Some CEPT countries obviously had in mind at that time that digital plan entries of GE06 could be used for mobile services like UMTS in the first place. The RR contain a footnote saying that in some countries part of UHF is allocated to mobile services on a co-primary basis. However, this was not valid for all CEPT countries. Extending the list was, however, not possible for the RRC-06. Only a World Radio Conference (WRC) can change the RR. Therefore, 53 countries signed Declaration 42, which basically states that the signing countries will make use of their digital plan entries also for services, which are not listed in the RR. The application of the envelope concept in that case, however, is subject to bilateral agreements.

It is difficult to say whether or not the decision of the RRC-06 to include the envelope concept in the GE06 Agreement had an impact on the consensus about the transition period. In any case, it was decided that it should end by 2015, except for some countries for whom the end of the transition period for VHF was fixed at 2020. It is, nevertheless, obvious that the envelope concept seemed to be very attractive to administrations in the sense that a lot of flexibility was incorporated into the GE06 Agreement, allowing administrations to adapt the usage of the spectrum to potential future demands.

8.2 Digital Dividend

As the RRC-06 ended, broadcasters thought the door would now be open to a new and prosperous future for broadcasting, but several administrations and operators of other telecommunication services had only waited for the RRC-06 to finish to reach out for new spectrum. The initiative that actually started already before the RRC-06, was based on the simple argument that digital terrestrial broadcasting allows a far more efficient usage of spectrum than analogue broadcasting. Consequently, if the existing analogue television services would be provided in terms of DVB-T, only 20–25% of UHF spectrum would still be required. The remaining could, in principle, be used by other systems, in the first place by mobile telecommunication services like UMTS. Therefore, immediately after the RRC-06 discussions started off within all relevant international and national organizations and entities concerned about frequency management.

The spearhead of these activities was the European Commission (EC), in particular its Radio Spectrum Policy Group (RPSG) (see Section 3.3). In its "Opinion on EU Spectrum Policy Implications of the Digital Dividend" [RSP07] a definition of the so-called Digital Dividend (DD) was given, which is usually referred to in Europe. It says that "Digital Dividend is to be understood as the spectrum made available over and above that required to accommodate the existing services in a digital form in VHF (Band III: 174–230 MHz) and UHF (Band IV and V: 470–862 MHz)." The DD is expected to be fully available throughout Europe, only after the end of the transition period, which in CEPT countries will be 2012.

In principle, the definition of DD, which only refers to Bands III, IV, and V could extended to include Band I (47–68 MHz) as well. Even before a digital broadcasting plan existed, broadcasters where starting to fade out analogue TV transmissions in Band I, because in that frequency band more man-made noise has to be dealt with than in other bands. After the digital switch-over, it can be expected that Band I will be fully cleared from broadcasting services. Therefore, this band could be seen as part of the DD as well.

There is no doubt that digital terrestrial broadcasting systems make use of the spectrum in a more efficient way than analogue television, for example. Actually, this was one of the arguments for the development and the introduction of T-DAB and DVB-T. Therefore, mapping all existing analogue TV services will certainly release a large amount of spectrum. The question only is for what purpose the newly available spectrum will be used.

The EC put forward three possible usages [ECo05]. The first possibility is to use the spectrum for the improvement of digital terrestrial broadcasting services. This refers to both technical quality of the delivered content, for example, High Definition Television (HDTV), as well as simply to offer more programs. But also enhancements or new features for existing programs could be envisaged. This could be multicamera angles for sports event transmissions or additional information, which is delivered on top of the programs such as special news streams.

The second potential usage of the DD is to devote spectrum for the introduction of multimedia broadcasting services targeting at mobile reception, such as DVB-H. This refers basically to a convergence between traditional broadcasting and individual mobile communication. These kinds of services bridge the gap between one-to-many broadcasting services and one-to-one individual mobile communication. Finally, the DD could be used by services, which are not connected to broadcasting at all. The next generations of mobile communication systems, such as UMTS or Long Term Evolution (LTE) [LTE08] could be candidates. Mobile broadband access to the Internet in scarcely populated areas in terms of WiMAX [WiM08] technologies is another application, which falls in that category.

Estimating the DD in quantitative terms is very difficult and even a delicate issue. It resembles very much the question of equitable access to the spectrum, which was a constant source of quarrel during the RRC-06. The question how equitable access to the spectrum can be guaranteed can also not be finally quantified. There are too many different aspects that have to be taken into account. Even worse, they usually do not apply to all involved parties on the same level. Sometimes it is said in the context of a conference that "equitable access has been granted if all participating administrations are equally unhappy."

In the case of DD, already the starting point from which its amount in quantitative terms is evaluated is not uniform across CEPT countries. It is true that the RRC-06 was guided by the principle to provide each country with the same number of nationwide coverages. Generally speaking, in Band III one DVB-T layer and three T-DAB layers were planned for in Europe. In UHF, the general planning goal was to obtain seven DVB-T layers. But care has to be taken here. The term layer refers to the possibility to provide a particular service, like a DVB-T multiplex throughout the national territory of a country. Whether this is accomplished with one single allotment or a set of allotments, which taken together add up to cover exactly the total area of a country is up to the decision of individual administrations and technical constraints in relation to network implementation. Many countries were planning on the basis of assignments only. In that case, it is not possible to determine the service area of a plan entry from the very start because the service area of an assignment depends on the impact of all co-channel plan entries. In order to be sure to close all potential gaps, many transmitters might need to be introduced in one layer. Therefore, a layer built from assignments might exhibit significant overlap of service areas.

At first, the layer concept is a pure geometrical concept. It just gives information about coverage of geographical areas. It does not say anything about the way this coverage is achieved. T-DAB and DVB-T offer a huge amount of different operation modes, which differ in relation to the data capacity, the robustness of the signal against unwanted signal contributions, and last but not least the reception mode. Even though it was agreed to harmonize the input requirements to the planning process of the RRC-06 with respect to these parameters to a certain level, still there were great variations across the planning area.

Apart from the pure technical aspects of the achievable DD there is also a political dimension.[12] Broadcasting is organized very differently

---

[12]It even seems that this is actually the crucial point. The possibility to devote part of the UHF spectrum to mobile services like UMTS has to be considered in the first place as a political and economical decision where technical arguments are not fully developed in a neutral way. Rather, they are exploited just to support a certain

across CEPT countries. This refers to the administrative level at which
broadcasting is regulated as well as the political ideas about the usage
of spectrum for public and private broadcasters. In one country, only
national networks are important, while in other countries the focus is on
regional broadcasting due to special linguistic or cultural peculiarities.
There might also be a bias towards the usage of certain channels. Due
to historical or financial reasons, for example, the channels in the lower
UHF band might be used more extensively than those in the upper band
or vice versa. Taking all that together, it comes as no surprise that the
answers to the question "How large is the DD?" might be totally different
from country-to-country.

Evaluating the amount of spectrum that can be considered as part
of the DD is just one aspect. From the EC's point of view it is equally
important to push for a harmonization of the DD across Europe. This is
primarily market triggered, since for manufactures of any kind of communication devices it would be important to have a harmonized European
market in order to benefit from corresponding economies of scale. However, there are several obstacles that need to be overcome in order to
achieve a European harmonization of the DD. First of all, there still is
the transition period throughout which analogue television is granted
protection. This constraint will cease in Europe by 2012. Furthermore,
there are many other services, both having primary and secondary status,
which make use of the spectrum under consideration. Their spectrum
usage across Europe is nonuniform and quite often subject to individual
national conditions. Some of these services are crucial for the production
of broadcasting content like SAB/SAP (Services Ancillary to Broadcasting and Programme Making) and wireless microphones, which are very
extensively used on occasions of big cultural or sports event.

The RRC-06 set up a regulatory framework for digital terrestrial
broadcasting by establishing the GE06 plan, which constitutes a de facto
harmonization of spectrum usage across Europe. This is sufficient to exploit the DD for any future development of broadcasting. However, in
order to open the door for other services than broadcasting in UHF
further regulatory measures would be needed. In its opinion on the
DD [RSP07] RPSG proposes to consider the introduction of harmonized
subband(s) in UHF to be used by these services. It is emphasized that

---

point of view. At least this was sometimes the perception of the author of this book
when participating in several of the expert groups of CEPT and EU.

this should be done on a nonmandatory basis only, which would leave the final decision about the usage of the DD to national administrations. Nevertheless, making available a subband for non-broadcasting services will without doubt have an impact on the existing GE06 allocations. Together with the introduction of more advanced television coding and transmission systems such as MPEG-4 and DVB-T2 it is expected that the impact on GE06 could be minimized (see Chapter 2 for more details about MPEG-4 and DVB-T2). In particular, broadcasters are very skeptical about such proposals having in mind the millions of DVB-T receivers already in the European market, which would not be able to decode the MPEG-4 multiplexes.

8.3 Sharing Spectrum with Mobile Communication Services

After the end of the RRC-06, the European Commission started to prepare the grounds for the exploitation of the DD by non-broadcasting services. At the end of January 2007, the final version of the "Mandate to CEPT on Technical Considerations Regarding Harmonization Options for the Digital Dividend" was issued [ECo07]. Even though the RPSG opinion [RSP07] was still speaking about the possibility to further develop broadcasting, this was not addressed anymore in the mandate text. The EC considered its mandate as a first step "to explore the technical feasibility of relevant potentials uses of the future Digital Dividend." Potential problems should be identified in particular with respect to coexistence of different services in the same spectrum range and consequently possible spectrum management strategies should be assessed. As expressed in two Communications to the European Parliament the European Commission wanted to establish a framework within which the DD could be exploited in an optimized manner. Since this suits perfectly their needs, mobile operators supported and still fully support the EC's activities while broadcasters and also several CEPT administrations were and still are not in favor of the EC's approach.

In detail CEPT was asked to analyze several technical and regulatory issues and to prepare corresponding reports. Planning of DVB-T networks in UHF during the RRC-06 was based on three RPCs, which were to mimic typical broadcasting network structures. Traditionally, broadcasting networks are implemented with the help of a limited number of high-power station. Even in the era of SFN networks, high-power stations usually build the backbone of any broadcasting network. If necessary they are supplemented by low- or medium-power stations in order to fill local-or regional-coverage gaps. The compatibility analysis of the RRC-06 implies that the mutual interference of networks of similar structures, namely RPC 1, 2, or 3, is to be evaluated.

Mobile communication networks are usually based on low-power networks, where base stations only cover a very small area compared to a typical broadcasting coverage area. Therefore, these networks are in need of many transmitters. If broadcasting and mobile services have to share the same spectrum range, there will be high-power networks based on a small number of transmitters and low-power networks consisting of many transmitters. Hence, a new interference situation will appear. CEPT was asked to look into that and carry out appropriate studies, allowing to conclude under which conditions coexistence of high-power and low-power networks is feasible.

As already put forward in the RPSG opinion, the EC considered harmonization of subbands an appropriate approach to accommodate non-broadcasting services in UHF. Therefore, CEPT should investigate the possibility of harmonizing at EU level a subband for multimedia application and for mobile services. Multimedia applications are understood as "broadcasting-like" services such as, for example, DVB-H, T-DMB, or similar applications. Basically, they consist of a broadband downlink that might—but need not in any case—be accompanied by a smallband uplink in order to allow for a certain level of interactivity.

Frequency planning for digital terrestrial broadcasting has to make sure that co-channel networks are separated far enough from each other so that the mutual interference they impose onto each other is not harmful. Clearly, this means that between the boundaries of the service areas of co-channel, high-power networks the considered frequency is not used. However, quite often it is, nevertheless, possible to make use of this channel even inside this buffer zone for low-power transmissions, using specially designed antenna patterns. Actually, this degree of freedom was heavily exploited in the ST61 plan for analogue television. It started with roughly 5000 plan entries in 1961, and just before revision of ST61 at the RRC-06 more than 80,000 entries were contained in the plan.

During the RRC-06, it was stated that the number of requirements exceeded the capacity of the available spectrum bands. The results of the compatibility analysis revealed a huge number of mutually incompatible input requirements, which consequently could not all be satisfied by the plan synthesis process. Therefore, millions of administrative declarations overruling the results of the compatibility analysis were provided by administrations in order to get the requirements into the plan. Nevertheless, it has to be noted that the planning was mainly based on RPC-RN combinations, which represent high-power networks. Also, in that case it was clear from the very beginning that there will still be (limited) room to introduce further low-power stations without causing too much interference. Broadcasters being aware of this naturally considered that as a possibility for future development of the broadcasting services. However, the EC in its mandate to the CEPT was asking the question whether it would be possible to use these so-called white spots between allotments for new or future applications and services.

CEPT was expected to deliver three reports covering these three issues. To this end, ECC Task Group 4 (ECC-TG4) was launched to prepare the demanded reports. Between January 2007 and April 2008 TG4 held 8 meetings in which the reports were drafted. Apart from administrations there were also participants from broadcasters and mobile operators in Europe. It comes as no surprise that the discussions in TG4 were sometimes very heated and consensus was very difficult to achieve. In the end, TG4 concluded that the coexistence of high-power broadcasting and low-power dense networks for multimedia applications can be managed by proper network design. For multimedia applications a dedicated subband would not be necessary. However, in order to guarantee the coexistence of mobile services and broadcasting a separation of the UHF spectrum into dedicated subbands, one for broadcasting and one for mobile services would be mandatory. A subband for mobile services consisting at least of UHF channels 62–69 was proposed.

Proposing the creation of a subband for mobile services in UHF and giving arguments that it might be feasible not to interfere with broadcasting transmission in channels, 60 and below is just the first step. However, not in all CEPT countries there was a primary allocation of mobile services in the channels 62–69 in the Table of Frequency Allocations of the Radio Regulations of ITU. In some countries such an allocation was effective by means of footnote 5.316 of the RR for quite some time. Basically, this is the same situation as already encountered in relation to the envelope concept included in the GE06 Agreement. During the RRC-06, it became clear that a plan entry of GE06 could be used for another service only if there would be a corresponding primary allocation in the RR.

In the case of § 5.1.3 of the GE06 Agreement CEPT administrations issued Declaration 42 in order to resolve the issue on a bilateral basis.
The results of TG4, however, could be put into practice only if the RR
would be changed by a World Radiocommunication Conference (WRC).
In autumn 2007, the WRC-07 [ITU07b] was held in Geneva and one of
the items on the agenda was addressing exactly such a change.

Even though the agenda of WRC-07 contained a lot different and
contended topics, Item 1.4 relating to an allocation of mobile services
in the UHF band kept the conference busy almost till the end. In particular, Europe and America kept struggling to overcome their dissent.
But finally, consensus could be reached on the basis of accepting different solutions for different regions of ITU. For Europe this means that
the frequency range between 790–862 MHz will be available for mobile
services on the basis of a co-primary allocation with broadcasting. This
means that current and future broadcasting services will have to share
this part of UHF with mobile services. The allocation will be effective
from June 17, 2015, which corresponds to the end of the transition period of the GE06 Agreement. This decision recognizes that during the
analogue–digital switch-over this spectrum will be heavily used and very
likely there will be little chance for free spectrum to mobile services.
Furthermore, any operation of mobile services will be subject to successful application of the GE06 procedures given in Article 4 and 5. From
a broadcaster's point of view this is very important, since it safeguards
protection of broadcasting services according to GE06.

Before the WRC-07 there was already footnote 5.316 of the RR,
which gave primary allocation to mobile services in the band 790–862
MHz in those countries listed in the footnote. WRC-07 decided to limit
the validity of footnote 5.316 till June 16, 2015 in order to match it
to the new harmonized allocation after that date. In Europe, footnote
5.316 covers Germany, Denmark, Finland, United Kingdom, and Netherlands just to mention some. Basically, this means that mobile services
could be introduced before 2015 in the listed countries. However, the
fact that there are many countries in Europe, which are not included
in footnote 5.316 (like Ireland, Italy, Belgium, Hungary, and so forth)
means that the protection of broadcasting services imposes severe restrictions for the operation of mobile services before 2015. There were
several countries at the WCR-07 who wanted to join the list of countries
of footnote 5.316. However, due to formal reasons this was not possible
and hence an additional footnote, namely 5.316A was included allowing

an earlier introduction of mobile services before 2015 for other countries as well.

Changing the Table of Frequency Allocation of the RR to allow mobile services in UHF channels 61–69 needs to be based on a proper assessment of the corresponding technical sharing criteria. To this end, WRC-07 issued Resolution 224 in order to clarify any open issues in relation to the introduction of mobile services in that spectrum band (among others). In particular, Study Groups of ITU were asked to study the impact on the GE06 Plan, as updated, and its future developments. ITU-R Recommendations are to be developed on how to protect the services to which the frequency bands under consideration are currently allocated.

Broadcasters struggled heavily to avoid a co-primary allocation in UHF for mobile services. In the end, WRC-07 decided to open channels 61–69 for these services. The technical conditions under which sharing of the spectrum can be achieved will be clarified until the WRC-11. Whether administrations will indeed give licenses for UMTS or other mobile communication services and to what extend is yet to be awaited. However, what is clear is that the generation of a harmonized subband in UHF across Europe would without doubt have severe impact on the GE06 plan.

The usage of channels 61–69 for broadcasting according to GE06 is very different across Europe. There are countries which do not have any allocations there at all while others make extensive usage of that spectrum range in order to provide several national layers of broadcasting services. In case administrations decide to free channels 61–69 from broadcasting plan entries, the question arises what will be done with the affected plan entries. Basically, there are three possibilities to deal with that problem. First, an administration could simply accept the loss of broadcasting plan entries in GE06 due to the fact there is no need for the seven layers as provided by GE06. If only few plan entries are affected it might be feasible to try to find alternative channels for those entries on a regional or even local basis in terms of an Article 4 procedure of the GE06 Agreement. In case this is not possible and there would be many countries having the need to reconstitute the previous broadcasting layers, then a full replanning for digital terrestrial broadcasting in the remaining frequency range channels 21–60 might be the only solution. Whatever will be felt necessary, it seems to be clear that major efforts are ahead in order to provide sufficient terrestrial broadcasting services in the future.

8.4   The Future of Terrestrial Broadcasting

The European Commission was one of the driving forces behind the coprimary allocation process for mobile services by putting forward a purely market oriented approach to spectrum management as has been described in a Communication to the European Parliament [ECo05a]. Basically, the idea is to modify the existing regulatory framework under the custody of ITU and CEPT and the national administrations towards a spectrum management approach, which will only follow supply and demand mechanisms. The introduction of a subband in UHF for mobile services is just the first step on this way. However, it seems this is considered as not sufficient in view of the explanations given in another Communication issued in 2007 [ECo07a]. Three subbands in the UHF range are proposed, one for traditional terrestrial broadcasting based on high-power networks, a subband for medium-power networks delivering multimedia content, and finally the subband for mobile services. From a regulatory point of view only the broadcasting part should remain in the realm of national administrations while the two others are partially or fully under the authority of the EC. The role of the Commission is justified by the claim that only a pan-European market driven spectrum management would allow efficient use of the spectrum.

Whether such an approach is appropriate or not or whether it is true at all is under heavy discussions in Europe. However, one thing seems to be obvious. Digitization of telecommunications and broadcasting services and their delivery has qualitatively changed the situation. In the case of terrestrial broadcasting, the production of content and the operation of the networks was in the hand of broadcasters in analogue times at least at the beginning. Emerging technologies and the advent of competitors both concerning content as well as network operation removed the competitive edge of broadcasters. Nowadays they have to act as any other commercial organization in the market. However, for public broadcasters this constitutes a major problem. Public broadcasting cannot be seen from a market oriented point of view only. It is always subject to many political constraints. In addition to market forces, they define in the first line the confines to which public broadcasters are bound.

The combination of politics and broadcasting differs drastically from country-to-country. In addition, the position public broadcasters have in economic terms also shows great variations across Europe. Without doubt terrestrial broadcasting generates a significant public value for society. Some broadcasters are committed to providing certain content of educational, informational, and cultural character targeting at special audiences due to political will and decision. This makes it extremely difficult to apply a strictly market-driven approach to spectrum management on a pan-European level.

On the other hand, in some countries in Europe terrestrial broadcasting is only of subordinate importance due to high cable and satellite penetration. Germany is a good example thereto. Averaged over the entire country, roughly 8% of the households referred to terrestrial television as primary reception mode before the digital switch-over. The figures were higher in large cities like Berlin or Munich but not all are comparable to United Kingdom, were the average was about 60% or more.

Both cable and satellite provide a large number of TV programs. Even if all seven layers planned for at the RRC-06 would be used, a maximum of about 28 programs would probably not be exceeded, depending on the system variant of DVB-T. Compared to the hundreds of programs in cable or on satellite, this is certainly a disadvantage. Clearly, if more efficient coding schemes are employed this disparity would be better balanced. What is important to notice, however, is that both cable and satellite are not able to provide portable or mobile coverage and probably this is the direction which is to be focused on for digital terrestrial broadcasting in the future.

This is also connected to the developments in relation to DVB-H and T-DMB. In both cases, traditional broadcast content is enhanced by offering additional services, which can be basically any kind of multimedia content. If this is incorporated into a mobile communication device a component of interactivity could be added by providing asymmetrical services with broadband downlink channel via DVB-H or T-DMB and small band uplink on the basis of the GSM or UMTS. Other systems like LTE were designed to allow for a hybrid operation between individual communication and broadcasting. As discussed, all these systems quite naturally would stimulate the erection of dense networks of low-power transmitters. Traditional high-power networks are very likely to become dispensable. New trends like femtocell technology [Fem08] or self-organizing networks [Fra08] might even accelerate the decay of traditional high-power broadcasting networks.

Apart from the delivery mechanisms radio and television already today see a dramatic change relating to the requested content. Historically, programs were broadcast according to a scheme the broadcasting companies decided themselves upon. Nowadays, the technical means are available to let the costumer have more freedom in deciding how and when a certain content is watched or listened to. Streaming on the Internet and podcasting constitute new forms of radio and video content. Together with hard drive recording devices they give costumers all possibilities to consume content perfectly tailored to their needs. Whether the programs are consumed in the future in the same way as they were for ages or not, there will certainly remain a strong demand for radio and television content in the future. The way the corresponding programs will be delivered in the future is an open question and it will certainly be interesting to follow the developments in this field.

Appendix A

Stochastic Optimization

Both frequency assignment and network planning fall into a class of problems which on a first glance seem to be trivial but prove to be highly complex and involved on second consideration. The crucial point in both cases is the fact that there exists a gigantic number of potential solutions. In order to assess a single solution, usually, a number of, sometimes, complicated calculations has to be carried out. Thus, it is prohibitive even in the simplest cases to simply try to assess all existing solution sequentially in order to find the optimum.

A simple example taken from the field of frequency assignment is to illustrate this. It is assumed that there is a set of M geographically distinct areas, which are to be allocated a TV channel each. However, this cannot be done freely. Instead, there is the constraint to be obeyed that geographically adjacent areas are not allowed to use the same channel. The total number of available channels should be N. Each of them should be usable in any area.

Clearly, a channel assignment onto the areas should be established which employs as few channels as possible. Thus, there is a variety of N^M different channel distributions. The only information available based on which an assessment of a frequency assignment could be carried out are the adjacency relations between areas. This, however, is not enough to exclude certain solutions a priori. Hence, in principle, all N^M possible configurations would need to be checked in order to find the best solution.

Now, if N and M are fixed to $N = 20$ and $M = 10$, for example, then there are 20^{10} different configurations. If the assessment of a single solution takes only 10^{-6} s, the exhaustive search for the best solution would require $2845\,h$ or 118 days! Since it is well known that typical frequency assignment problems dealt with at international radio conferences must cope with several hundreds of coverage areas to be assigned frequencies to, it becomes clear that even with the help of the fastest today existing computers such an approach must fail.

A very similar situation is encountered in the case of network planning. Typical wide-area T-DAB or DVB-T networks encompass a set of 10–50 individual transmitters that are operated as a single frequency network. Each of them is characterized by fixing the set of network parameters such as the radiated power, its antenna diagram, and the individual broadcasting time delay, which might be used to synchronize the whole network and to reduce self-interference.

Usually, antenna diagrams are defined in terms of 36 numbers, which correspond to a reduction of the effective radiated power towards a given radial direction. The standard of 36 values simply stems from the fact to use a $10°$ step between two directions. Therefore, every transmitter in that simple model is described by a set of $1 + 36 + 1 = 38$ parameters.

In principle, these parameters can take any real value. However, practically only distinct values taken from a given set are used for planning purposes. For example, both ERP and antenna diagrams are usually given in terms of multiples of 1dB within certain limits while for the time delays the basic unit is $1\,\mu$s. Now, assuming for all parameters a 10-member set, each transmitter can be configured in 10^{38} different ways. If the single frequency network is build by a medium number of transmitters, let's say 30, there are $30 \times 10^{38} = 3 \times 10^{39}$ possible configurations of the network as a whole. As before, any brute-force approach to find the best solution must necessarily fail.

From a mathematical point of view, both planning problems just described fall into the class of NP-hard problems. This characterizes a task whose systematic solution by testing every existing possibility leads to computing times, which grow exponentially or even faster with respect to the relevant system size, that is the number of individual objects or parameters. In the case of typical system sizes encountered in real-world problems very quickly orders of magnitude are reached that can no longer be tackled that way.

In view of these well-known difficulties, there has been put forward a tremendous effort during the last three decades in order to develop powerful methods, which allow at least for a very efficient approximative treatment of the underlying high-dimensional optimization problems. They have been successfully applied to frequency assignment or network planning problems. These methods are the so-called stochastic optimization algorithms.

Stochastic optimization algorithms come in several different flavors. The core feature of all different approaches is to carry out a stochastic search in configuration space, which in contrast to pure Monte-Carlo methods is accompanied by mechanisms that more or less effectively guide the search towards the desired solution. What is very important, however, is the fact that in any case, the basic idea consists of not seeking the one and only global optimal solution, but to accept a "nearly optimal" solution as well. The great advantage is that such a solution can be found within a reasonable period of time.

Meanwhile, there is a vast number of different stochastic optimization algorithms. Some of them show only marginal differences. Therefore, only two approaches are described here in detail which have been exploited in the field of frequency and network planning. For further examples it is referred to the literature.

A.1 Great Deluge Algorithm

A special representative of stochastic optimization algorithms is the so-called Great-Deluge Algorithm (GDA) [Due93]. Even though it is amazingly simple it produces very good and robust results. Its applicability is restricted by very few conditions only. Typical fields of application are problems in which a system is described in terms of a large set of parameters, which determine the quality of the system in a very complicated way. Examples can be found in physics or economics when the degree of efficiency of a technical system or some logistic scheme has to be optimized.

The most important condition that has to be met in order to apply the GDA to a particular optimization problem is that the status of the system under consideration can be fully described by fixing a set of parameters. Further, it must be possible to quantify the status of the system in a non-ambiguous way. This means a quality function must exist, which maps the set parameters to a reasonable quality measure. Finally, it is required that good configurations can be distinguished from bad ones and that the difference can be quantified numerically as well.

There are not many restrictions on the explicit form of the quality function neither on the definition nor on the parameters. The domain of

the parameters can be the set of integers or real numbers but in principle complex number are conceivable as well. The parameters might be
restricted to certain intervals or even to a discrete set of values. The
same is true for the quality function. It can be a continuos real-valued
function or an integer function. Only complex-valued quality functions
would create problems, because in that case there is no clear mathematical distinction between good and bad.

Seen from a mathematical point of view the set of parameters spans a
high-dimensional configuration space. The quality function can be visualized in the sense that it constitutes a very complex mountainous region
above that configuration space as a function of its parameters. Depending on the domain of the parameters the quality function need not be
a continuos more-dimensional surface. If the set of parameters contains
both real and discrete values, then the quality function necessarily has
discrete or continuos character in different directions of the configuration
space.

The mathematical task underlying the optimization is to find a set
of parameters that corresponds to a state or configuration of the system having optimal quality. In the case of the Great Deluge algorithm,
this is accomplished in an iterative manner based on a stochastic search
in configuration space. Decisive for the success of the method is that
the accessibility of certain regions in configuration space is dynamically
reduced, that is more and more sectors become inaccessible as the simulation proceeds.

The course of the GDA can be summarized as follows. Starting point
of the optimization process is an arbitrarily chosen configuration whose
quality is calculated first. Then a new configuration is generated and
its quality calculated as well. According to a given scheme the new
configuration is either accepted as the new starting point or not. This
procedure is continued until a particular stop criterion for the simulation
is met.

Basically, this whole process is governed by two factors. First, the
newly generated configurations should be in the "vicinity" of the old
ones. This presumes that a vicinity of configurations can be defined at
all. In most cases, an adjacent configuration can be generated by slightly
changing the values of the parameters of the configuration at hand.
Then, the quality of the new configuration will change only slightly, too.
In the case of real parameters and smooth quality functions this is obvious. However, in general this might be far from being trivial. Depending

on the structure and the complexity of the quality function it might be not possible at all.

Second, the decision which configurations are accepted as new starting points is taken dynamically during the simulation. The decision criterion applied is if the quality of the newly generated configuration is better than a dynamically increasing lower bound or not, the so-called water level WL. If the new quality is better than WL, then the new configuration is accepted as new starting point even if its quality is less than the old quality. Thus, degradations are temporarily accepted. Only if the new quality is smaller than WL the corresponding configuration is rejected and a new one has to be tried out. Depending on the change in quality when going from one configuration to another the water level is increased after each step. This way the dynamic blocking of certain parts of configuration space is realized.

The name "water level" stems from the way in which the optimization process can be visualized and which finally has been the origin of the name "Great Deluge" algorithm. The situation is similar to a blind man trying to find his way up to highest mountain peak or at least to a very high one. His situation is aggravated by the fact that on his way upward it starts to rain so heavily that the valleys start to fill. Great Deluge has finally started. Furthermore, he is barefoot, which certainly should be avoided in high mountains.

Since the man does not see anything he can only make one small step after the other. So, from where he actually stands he makes one single move into an arbitrary direction. If his feet become wet he immediately goes back to his former position and tries another step. If ending up on dry land the man accepts his new position. While the water level is permanently rising the man is pushed up the hill towards a high peak. For the man's sake the rain should stop once he arrives at a peak from which there is no escape anymore due to being encircled by water. In mathematical terms, this corresponds to the stop criterion. Figure A.1 illustrates the process in terms of a one-dimensional example.

At the beginning of the simulation, the initial configuration x_0 is chosen and the corresponding quality f_0 is calculated. Furthermore, the water level WL is fixed. The first iteration step consists of slightly changing the value x_0 and thus position x_1 is reached. Its quality reads f_1 which in example given here is smaller than that of x_0 but larger than WL. Thus, it is accepted as the new starting point. Next, the water level is raised.

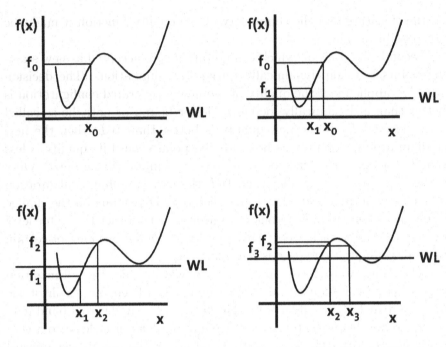

Figure A.1: One-dimensional example for the Great Deluge algorithm.

During the second step the location x_2 is reached. If during the attempt to leave point x_1 a value x_2 is chosen, which leads to quality smaller than WL, then this it would be necessary to reject this x_2 and try to find a better one. Repeating this procedures a sequence of points x_k is tested and finally a good solution can be found. The crucial point, however, is that by accepting temporarily reduced qualities, it is possible to escape from the local maxima as is illustrated by the third step in Figure A.1. Thus, the path is open to reach the global maximum in this case. Figure A.2 shows a flow chart of the GDA.

A.2   Evolutionary Strategies

Besides all algorithms that have their origin in different mathematical considerations there is one omnipresent, very flexible, and overall very successful strategy to cope with high-dimensional optimization problems. Nature itself applies this strategy, which is evolution. On the basis of recombination, mutation, and selection, it has been possible to create an

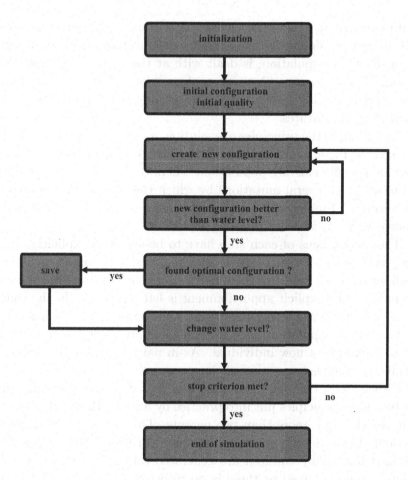

Figure A.2: Flow chart of the Great Deluge algorithm.

almost unbelievably huge biological diversity on earth. Creatures were able to adopt to sometimes extreme living conditions.

The application of the underlying principles to the solution of high-dimensional optimization problems is rather young discipline. At about the end of the 1980s of the last century the first attempts into that direction have been made. Since that time, there has been great progress both with respect to the development of new algorithms and the application to concrete practical problems. A vast number of articles has been produced and published. In order to get a first impression of the field it is referred to [Abl87] and [Gol89].

In contrast to the Great Deluge algorithm, evolutionary strategies (ES) do not consider only one current configuration. Instead, an entire set, a so-called population, is dealt with at the same time, that is during one simulation step. A single configuration, called an individual, is characterized by its "genes," which is nothing but a set of parameters quantifying its features.

According to the principles of evolution, an simulation step consists of three parts. First, a number of offspring have to be generated by starting from an actually existing population. Then, each of the new individuals has to undergo several mutations by which the values of its parameters are affected. Finally, a subset of the entire set of new individuals is selected which is to build the new population.

The mechanisms of each step have to be specified explicitly. In nature, parents pass theirs genes to their children. In general, a child gets a full set of genes. Some of them derive from the mother and some from the father. The explicit apportionment is left to chance. In the context of an evolutionary strategy this means that parents, that is individuals of the current population, pass their genes, that is their parameter values to a child or a new individual. As in nature, who is passing which parameter value is a question of chance as well.

Clearly, for a computer simulation there is no need to literally adopt the biological principles put into practice by nature. Hence, it is possible that one child has more than two parents. This is one of the degrees of freedom that can be used to adjust the algorithm. Also, it has to be specified how many children are generated during one iteration. This number could be fixed or there is no principle objection to let it float according to a particular scheme.

The next step, namely mutation, needs to be defined properly as well. The number of parameters to be changed and also the extent of change must be fixed. It is possible to apply a modification to all parameters or to only change some of them, however, in a drastic way.

Finally, the population has to be set up. Also this step allows for a large amount of freedom. The new population could be recruited exclusively from the set of children. It is, nevertheless, conceivable to build a new population on the basis of both parents and children. Furthermore, what is the criterion for an individual to be included into the new population? Is it only the corresponding quality, are the individuals selected completely by chance or is it a mixture of both?

From this discussion it becomes clear that in comparison to the GDA, evolutionary strategies bear a lot more degrees of freedom. This can be helpful in some cases in order to adapt the algorithm to a particular problem in a perfect way. However, on the other hand it is evident that a lot more effort has to be put into the application of the ES as with the other simple, GDA. Figure A.3 illustrates the basic ingredients of an evolutionary strategy in terms of a flow chart.

From a conceptional point of view, evolutionary strategies constitute some kind of generalization of the simple type of algorithms like Great Deluge. Not only one configuration is treated at one time but also a whole set of system states are dealt with and assessed during one iteration step. This demands a significantly larger administrative effort.

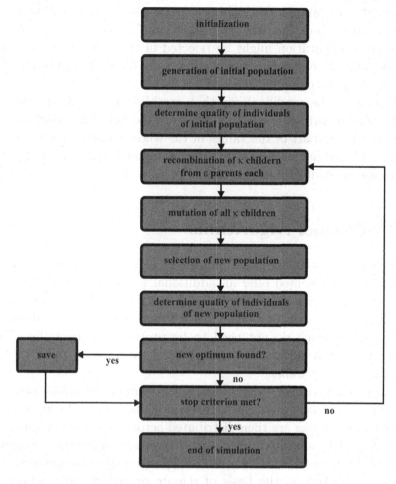

Figure A.3: Flow chart of an evolutionary strategy.

Correspondingly, the requirements in relation to computer resources like RAM are getting more severe, too. In addition, the computing times are increasing because not only one but several configurations have to be dealt with. On the other hand, the evolutionary algorithms are usually converging much faster towards an acceptable final solution.

The great difference between the ES and GDA lies in the selection and the acceptance of new configurations, respectively. Basically, in the case of the GDA there is a global criterion according to which it is decided if a new configuration is accepted or not. If the quality is larger than the water level the new configuration is taken independently from the difference in quality between the old and new configuration.

In the case of an evolutionary strategy a relative criterion is employed. If, for example, the 20 best configurations are supposed to build the new population, then the qualities certainly show some variation. A moderate configuration might be rejected in the context of one particular population, but could have a completely different significance with respect to another population.

Moreover, the modifications that are allowed in ESs usually lead much more drastic changes as is typically accepted in GDA-type approaches. Using part of the values of the whole parameter set from one parent and another part from another individual, quite naturally gives rise to discontinuities in terms of quality.

A.3   Parallel Algorithms

Apart from their efficiency and their robustness stochastic algorithms like the ones just presented offer an additional feature, which makes them very attractive from a computational point of view. Nearly all methods can be parallelized in different ways without problems. All algorithms posses parts, which do not need to be processed sequentially. Thus, quite naturally a performance gain can be expected, if these subtasks are processed in parallel.

There are two basic ways to carry out parallel computations. The traditional approach is to utilize several connected workstations. The computational tasks are then distributed across the local computer network onto the individual machines. Nowadays, combining computers across the Internet is also an option. The distributed computation usually is established on the basis of remote procedure calls, whereupon processes are started on remote machines in order to carry out some work.

If a particular computer is equipped with more than one processing units, then this fact can be exploited for parallel computing purposes as well. Multithread programing techniques are the right tool to make optimal use of this potential. For a detailed presentation of multithread programing it is referred to [Lew96] and [Kle96].

It is clear that a multitasking operating system is a prerequisite for any multithread programing activities. Multitasking operating systems can do more than one job at given time by running more than one process concurrently. In principle, a process can do the same thing by starting more than one thread at the same time. Each thread is a distinct program flow, which carries out individual programing statements. One thread, for example, can open a window for displaying some graphic on screen while a second is occupied with I/O tasks and the third is making some numerical calculations.

What is the fundamental difference between a process and a thread? A process is a unit belonging to the operating system. It is characterized by a virtual memory map, file descriptors, user-IDs, source code, and so on. The only possibility of a program to access the data of the process or to enforce some change of the status of the process is via a system call, which is a very expensive action where the phrase expensive refers to computer resources and computational time.

In contrast, a thread is a unit on user level, which can be manipulated by ordinary user functions. Thus, a process associated to particular executable code can generate independent entities, namely threads, in order to carry out dedicated tasks independent of any other concurrent operations. This structure allows the generation of threads on different CPUs of a single machine without the need for special hardware.

If a program consisting of just one single thread requires some task to be carried out by the operating system, the whole program must wait until the operating system has finished that task. Then the next program command can be executed. In contrast, a multithread program can execute other commands while waiting. Thereby, the total performance is increased and computing resources are utilized in a more efficient way. Performance is enhanced also by the fact that both the creation of threads as well as the interthread communication are much faster then the corresponding process operations. For a more detailed description of multithread programming techniques it is referred to [Lew96].

A closer look on Figure A.3 reveals a quite obvious way to parallelize the previously discussed ES algorithm. During each iteration step a prescribed number of children is generated and assessed. Usually, this is carried out sequentially. Hence, it would be possible to exploit the multithread programing technique in order to start a set of threads each of them having the task to generate and assess exactly one child. Once this has been done the program proceeds again along the traditional line and decides which individuals are to be included into the population.

When assessing a configuration, it is very often possible to subdivide this task into distinct pieces, which could dealt with independently at the same time. Such a situation is found, for example, when calculating a coverage prediction for a SFN (see Chapter 4). For every geographic point representing the coverage area the same set of calculations has to be carried out. Without problems the whole set of pixels could be subdivided into a number of smaller sets, each of them being dealt with by an individual thread, independent from the others. At the end the results of all threats are combined to give the total coverage.

Also for the Great Deluge algorithm parallelization is a conceivable way to improve the computational performance. The flow chart from Figure A.2 could be modified not only by creating one new configuration per iteration step but by allowing the creation of a multitude of them. Both creation and assessment of these configurations could be carried out concurrently in terms of independent threads. Then, the selection mechanism is adapted in the sense that either the best or the worst configuration just above the water level is chosen as the new starting point.

The question if parallelization is to be exploited or not and in which form, that is via multithread programing techniques or via remote procedure calls to other machines, cannot be answered generally. This depends very heavily on the details on the considered optimization problem. In some cases, parallelization is a good idea in others it is of no help because, for example the administrative overload exceeds the potential performance gain.

In connection with the planning of digital single frequency networks, investigations have been carried out to study the impact of multithread programing techniques on computational performance [Beu98b]. To this end, the coverage has been calculated for a given network within a prescribed planning area with the help of threads the way described above on a multiprocessor machine. It has been expected that a significant performance gain could be achieved. However, the results seem to dampen the enthusiasm a little bit. Unfortunately, the simple equation N CPUs = N threads = N times faster is not valid. Increasing the number of threads inevitably increases the administrative effort as well, which obviously limits the performance gain. Nevertheless, an improvement of about $N/3$ times faster, where N is the number of CPUs, can be realized without problems.

Appendix B

Abbreviations and Symbols

The following abbreviations are used throughout the book:

ABA	:	African Broadcasting Area
BR	:	Radiocommunication Bureau of the ITU
CEPT	:	Conférence Européenne des Administrations des Postes et des Télécommunication
CERP	:	European Committee on Postal Regulation
CH97	:	The Chester Multilateral Agreement relating to DVB-T, Chester, 1997
CNG	:	Coordination and Negotiation Group
COFDM	:	Coded Orthogonal Frequency Division Multiplex
CPG	:	Conference Preparatory Group
CPM	:	Conference Preparatory Meeting
DAB, T-DAB	:	Digital Audio Broadcasting
DD	:	Digital Dividend
DQPSK	:	Differential Quadrature Phase Shift Keying
DRM	:	Digitale Radio Mondiale
DVB, DVB-T	:	Digital Video Broadcasting
EBA	:	European Broadcasting Area

EBU	:	European Broadcasting Union
EC	:	European Commission
ECA	:	European Common Allocation Online Database
EP	:	European Parliament
ERC	:	European Radiocommunications Committee
ERO	:	European Radiocommunication Office
ERP	:	Effective Radiated Power
FFT	:	Fast Fourier Transform
FIC	:	Fast Information Channel
FM	:	Frequency Modulation
GE06	:	Final Acts of the Regional Radiocommunication Conference, Geneva, 2006
GE89	:	Final Acts of the Regional Administrative Radio Conference, Geneva, 1989
GSM	:	Global System for Mobile Communication (formerly: Group Spécial Mobile)
HDTV	:	High Definition Television
IPG	:	Intersessional Planning Group
ITU	:	International Telecommunication Union
MA02	:	CEPT T-DAB Planning Meeting Maastricht (4), 2002
MIFR	:	Master International Frequency Register
MPEG	:	Moving Picture Expert Group
LNM	:	Log-Normal Method
PAL	:	Phase Alternating Line
PXT	:	Planning eXpert Group
QAM	:	Quadrature Amplitude Modulation
QPSK	:	Quadrature Phase Shift Keying

RA	:	Radiocommunication Assembly
RPG	:	Regulatory and Procedural Group
RRC-04	:	First Session of the Regional Radiocomminications Conference, Geneva, 2004
RRC-06	:	Second Session of the Regional Radiocomminications Conference, Geneva, 2006
RSC	:	Radio Spectrum Committee of the European Commission
RSD	:	Radio Spectrum Decision
RSPG	:	Radio Spectrum Policy Group of the European Commission
SAB/SAP	:	Services Ancillary to Broadcasting and Program Making
SRD	:	Short Range Devices
ST61	:	Final Acts of the European VHF/UHF Broadcasting Conference, Stockholm, 1961
TFA	:	Table of Frequency Allocations
TG 6/8	:	Task Group 6/8 of the ITU
UHF	:	Ultra High Frequency
UMTS	:	Universal Mobile Telecommunications System
UN	:	United Nations
UPU	:	Universal Postal Union
VHF	:	Very High Frequency
WI95	:	CEPT T-DAB Planning Meeting Wiesbaden, 1995
WRC-07	:	World Radio Conference 2007

The following table contains the explanations of the most important mathematical parameters that are used throughout this book:

A_{km}	:	antenna diagram factor of the k-th transmitter for the m-th direction
E_l	:	number of inhabitants corresponding to pixel l
F_{min}	:	minimum field strength
F_{total}	:	total field strength
F_{wanted}	:	wanted field strength
$F_{unwanted}$	:	unwanted field strength
F_P	:	protection ratio
G_d	:	antenna gain
K_k	:	costs of k-th transmitter
P_k	:	ERP of k-th transmitter
Φ_k	:	direction of main lobe of k-th antenna diagram
R	:	reuse distance
T_G	:	guard interval
T_S	:	symbol duration of a T-DAB or DVB-T symbol
T_W	:	duration of FFT evaluation window of the receiver
Q	:	quality function
c	:	velocity of light
κ_k	:	width of k-th antenna diagram
$\mathbf{r}_k$	:	location of k-th transmitter site
τ_k	:	time delay of k-th transmitter

Bibliography

[Abl87]   Ablay, P., *Optimieren mit Evolutionsstrategien*, Spektrum der Wissenschaft, **7** (1987) 104

[Bal82]   Balanis, C. A., *Antenna Theory - Analysis and Design*, John Wiley & Sons, New York, 1982

[Beu95]   Beutler, R., *Optimization of Digital Single Frequency Networks*, Frequenz **49** (1995) 245

[Beu98a]   Beutler, R., *A-Priori- und A-Posteriori-Dibitfehlerwahrscheinlichkeit DQPSK-modulierter Übertragung über Fading-Kanäle*, Forschungsverbund Medientechnik Südwest, Tagungsband des Projekttages "Messung und Modellierung von Funkkanälen", Beutler, R., Prosch, Th. (Eds.), Südwestrundfunk, Stuttgart, 1998

[Beu98b]   Beutler, R., *Digital Single Frequency Networks: Improving Optimization Strategies by Parallel Computing*, Frequenz **52** (1998) 90

[Beu01]   Beutler, R., *Abschätzung des Spektrumsbedarfs für DAB*, telekom praxis **78** (2001) 26

The list of references given here has to be considered as a subjective subset of those documents that might be relevant. It is by no means exaustive. Rather, it reflects the author's knowledge and usage of literature. It might give first indications and hints for further reading. Additional references can be found with the help of any Internet search engine without problems.

Some of the references given point to documents that are not officially available. This concerns for example documents submitted to EBU project teams. However, in most cases these documents are not classified and can be obtained by addressing the EBU or the authors directly.

255

[Beu04a] Beutler, R., *FA - A Frequency Assignment Software Tool*, Südwestrundfunk, Stuttgart, Germany, 2004

[Beu04b] Beutler, R., *Frequency Assignment and Network Planning for Digital Terrestrial Broadcasting Systems*, Springer, NewYork, 2004

[Beu04c] Beutler, R., et al., *Ergebnisse des 1. Teils der Regionalen Planungskonferenz (RRC-04) zur Revision des Stockholmer Abkommens*, FKT, **11** (2004) 551

[Bre79] Brélaz, D., *New Methods to Color the Vertices of a Graph*, Communications of the ACM **22** (1979) 251

[Bru92] Brugger, R., *Die exakte Bestimmung der nutzbaren Feldstärke bei der Behandlung von Mehrfachstörungen und ein Vergleich älterer und neuer Näherungsmethoden*, Rundfunktechnische Mitteilungen **38** (1992) 23

[Bru05] Brugger, R., Meyer, K., *RRC-06 – Technical Basis and Planning Configurations for T-DAB and DVB-T*, EBU Technical Review, **302** (2005), Geneva, Switzerland

[Car90] Carraghan, R., Pardalos, P.M., *An Exact Algorithm for the Maximum Clique Problem*, Operations Research Letters **9** (1990) 375

[Cau82] Causebrook, J. H., et.al., *Computer Prediction of Field Strength - A Manual of Methods developed by the BBC for the LF, MF, VHF and UHF Bands*, Engineering Division, BBC, 1982

[CEN08] European Committee for Electrotechnical Standardization, www.cenelec.eu, 2008

[CEP95] CEPT, *Final Acts – CEPT T-DAB Planning Meeting, Wiesbaden, 1995*, European Conference of Postal and Telecommunications Administrations, Wiesbaden, Germany, 1995

[CEP96] CEPT, *Final Acts of the T-DAB Planning Meeing (2), Bonn, November 7–8, 1996*, European Conference of Postal and Telecommunications Administrations, Bonn, Germany, 1996

[CEP97] CEPT, *The Chester 1997 Multilateral Coordination Agreement relating to Technical Criteria, Coordination Principles and Procedures for the Introduction of Terrestrial Digital Video Broadcasting (DVB-T)*, European Conference of Postal and Telecommunications Administrations, Chester, United Kingdom, 1997

[CEP02] CEPT, *Final Acts of the T-DAB Planning Meeing (4), Maastricht, 2002*, European Conference of Postal and Telecommunications Administrations, Maastricht, The Netherlands, 2002

[CEP07] Conference Européenne des Administration des Postes et des Télécommunications, www.cept.org, 2007

[CEP07a] CEPT, *Final Acts of the CEPT Multi-lateral Meeting for the frequency band 1452–1479.5 MHz, Constanţa, 2007*, European Conference of Postal and Telecommunications Administrations, Constanţa, Romania, 2007

[CER08] European Organisation of Nuclear Research, www.cern.ch, 2008

[Cha07] Chari, M.R., et. al., *FLO Physical Layer: An Overview*, IEEE Transactions on Broadcasting **53** (2007) 145

[Com88] Commission of the European Communities, *COST 207. Digital Land Mobile Radio Communication – Final Report*, Brussels, Belgium, 1988

[Com05a] Commission of the European Communities, *Communication from the Commission to the Council and European Parliament: A forward-looking rdaio spectrum policy for the European Union; Second Annula Report*, COM (2005) 411

[Com05b] Commission of the European Communities, *Communication from the Commission to the Council, the European Parliament and the European Economic and Social Committee and the Committee of the Regions: A market-based approach to spetrum management in the European Union*, COM (2005) 400

[DRM07] Deutsches DRM-Forum, *Broadcasters' User Manual*, www.deutsches-drm-forum.de/Broadcast_ManualV2.pdf, 2007

[Due93]   Dueck, G., *The Great Deluge Algorithm and the Record-to-Record Travel*, J. Comp. Physics. **104** (1993) 86

[DVB08]   DVB Project, www.dvb.org/technology/dvbt2/, 2008

[EBU98]   European Broadcasting Union, *Technical Bases for T-DAB Services Network Planning and Compatibility with Existing Broadcasting Services*, EBU BPN Doc. 003-Rev.1, Geneva, Switzerland, 1998

[EBU01]   European Broadcasting Union, *Terrestrial Digital Television Planning and Implementation Considerations - Edition 3*, EBU BPN Doc. 005, Geneva, 2001

[EBU02]   European Broadcasting Union, *Planning Criteria for Mobile DVB-T*, EBU BPN Doc. 047, Geneva, Switzerland, 2002

[EBU03]   European Broadcasting Union, *Impact on Coverage of Inter-Symbol Interference and FFT Window Positioning in OFDM Receivers*, EBU BPN Doc. 059, Geneva, Switzerland, 2003

[EBU05]   European Broadcasting Union, *Guide on SFN Frequency Planning and Network Implementation with Regard to T-DAB and DVB-T*, EBU BPN Doc. 066, Geneva, Switzerland, 2005

[EBU07]   European Broadcasting Union, *Broadcasting Aspects Relating to the Procedures fo Coordination and Plan Conformity Agreement in the GE06 Agreement*, EBU BPN Doc. 083, Geneva, Switzerland, 2007

[EBU08]   European Broadcasting Union, www.ebu.ch, 2008

[ECo05]   European Commission, *Communication on EU Spectrum Policy Priorities for the Digital Switchover in the Context of the Upcoming ITU Regional Radiocommunication Conference 2006*, COM(2005) 461, Brussels, 2005

[ECo05a] European Commission, *Communication from the Commission to the Council, the European Parliament and the European Economic and Socail Committee and the Committee of the Regions – A Market-Based Approach to Spectrum Management in the European Union*, COM(2005) 400, Brussels, 2005

[ECo07] European Commission, *Mandate to CEPT on Technical Considerations Regarding Harmonisation Options for the Digital Dividend*, DG INFOSO/B4, Brussels, 2007

[ECo07a] European Commission, *Communication from the Commission to the Council, the European Parliament and the European Economic and Socail Committee and the Committee of the Regions – Reaping the Full Benefits of the Digital Dividend in Europe: A Common Approach to the Use of the Spectrum Released by the Digital Switchover*, COM(2007) 700, Brussels, 2007

[ERO07a] European Radiocommunications Office, www.ero.dk, 2007

[ERO07b] European Radiocommunications Office, *European Common Allocation Online Database*, apps.ero.dk/eca/, 2007

[ETS97a] European Telecommunications Standards Institute ETS 300 401, *Radio Broadcasting Systems: Digital Audio Broadcasting (DAB) to Mobile, Portable and Fixed Receivers*, 1997

[ETS97b] European Telecommunications Standards Institute ETS 300 744, *Digital Video Broadcasting (DVB): Framing Structure, Channel Coding and Modulation for Digital Terrestrial Television (DVB-T)*, 1997

[ETS04a] European Telecommunications Standards Institute EN 302 304 V1.1.1, *Digital Video Broadcasting (DVB): Transmission System for Handheld Terminals (DVB-H)*, 2004

[ETS04b] European Telecommunications Standards Institute EN 301 192 V1.4.1, *Digital Video Broadcasting (DVB): DVB Specification for Data Broadcasting*, 2004

[ETS05] European Telecommunications Standards Institute ES 201 980 V2.2.1, *Digital Radio Mondiale; System Specification*, 2005

[ETS07] European Telecommunications Standards Institute TS 102 563 V1.1.1, *Digital Audio Broadcasting: Transport of Advanced Audio Coding (AAC) Audio*, 2007

[ETS08] European Telecommunications Standards Institute, www.etsi.org, 2008

[EUR96] Eureka 147 Projekt, *Digital Audio Broadcasting*, 1996 [available from von Dr. Thomas Lauterbach c/o Robert Bosch GmbH, Abt. FV/SLM, Robert-Bosch-Straße 200, 31139 Hildesheim]

[Far06] Faria, G., et. al., *DVB-H: Digital Broadcast Services to Handheld Devices*, Proceedings of the IEEE, **94** (2006) 194

[Fem08] Femto Forum, www.femtoforum.org/femto/, 2007

[Fra08] Frauenhofer Institut für Integrierte Schaltungen, *Self Organizing Networks and Sensor Networks*, www.iis.fraunhofer.de/EN/abt/kom/son.jsp, 2008

[Gra02] Gräf, A., McKenney, T., *Ensemble planning for digital audio broadcasting*, in "Handbook of Wireless Networks and Mobile Computing", Stojmenovic, J. (Hrg.), John Wiley and Sons, New York, USA, 2002

[Gol89] Goldberg, D. E., *Genetic Algorithms in Search, Optimization, and Machine Learning*, Addison-Wesley, Reading, USA, 1989

[Gro86] Großkopf, R., *Feldstärkevorhersgae im VHF-Bereich mit Hilfe topographischer Daten*, Rundfunktechnische Mitteilungen **30** (1986) 176

[Gro95] Großkopf, R., *Vergleich verschiedener Feldstärkeprognoseverfahren mit Messungen im DAB-Gleichwellennetz in Bayern*, Rundfunktechnische Mitteilungen **39** (1995) 102

[Hal96] Hall, M. P. M., Barcley, L.W., Hewitt, M. T. (Hrgs.), *Propagation of Radio Waves*, The Institution of Electrical Engineers, London, UK, 1996

[Hes98] Hess, G. C., *Handbook of Land-Mobile Radio System Coverage*, Artech House Books, London, UK, 1998

[Hun96] Hunt, K., et. al., *The CEPT Planning Meeting Wiesbaden, July 1995*, EBU Technical Review, **267** (1996), Geneva, Switzerland

[Hun02] Hunt, K., O'Leary, T., Ratkaj, D., *Reflections on a Near-Past T-DAB Conference*, EBU Technical Review, **292** (2002), Geneva, Switzerland

[IMT00] International Mobile Telecommunications 2000, www.itu.int/home/imt.html, 2000

[IRT82] Institut für Rundfunktechnik, *Richtlinie für die Beurteilung der UKW-Hörfunkversorgung (Mono und Stereo) bei ARD und DBP*, München, Germany, 1982

[ISO93] International Organisation for Standardization, ISO/IEC 11172-1:1993, *Information technology – Coding of moving pictures and associated audio for digital storage media at up to about 1,5 Mbit/s – Part 3: Audio*, 1993

[ISO05a] International Organisation for Standardization, ISO/IEC 14496-3:2005, *Information technology – Coding of audio-visual objects – Part 3: Audio*, 2005

[ISO05b] International Organisation for Standardization, ISO/IEC 14496-10:2005, *Information technology – Coding of audio-visual objects – Part 10: Advanced Video Coding*, 2005

[ITU61] International Telecommunications Union, *Final Acts of the European VHF/UHF Broadcasting Conference*, Stockholm, Sweden, 1961

[ITU89] International Telecommunications Union, *Final Acts of the Regional Administrative Radio Conference for the Planning of VHF/UHF Television Broadcasting in the African Broadcasting Area and Neighbouring Countries*, Geneva, Switzerland, 1989

[ITU98] International Telecommunications Union, ITU-R Recommendation BS. 412-9, *Planning Standards for Terrestrial FM Sound Broadcasting at VHF*, Geneva, Switzerland, 1998

[ITU00] International Telecommunications Union, ITU-R Recommendation BT. 1368-2, *Planning Criteria for Digital Terrestrial Television Services in the VHF/UHF Bands*, Geneva, Switzerland, 2000

[ITU01a] International Telecommunications Union, ITU-R Recommendation P. 1546, *Method for Point-to-Area Predictions for Terrestrial Services in the Frequency Range 30 to 3 000 MHz*, Geneva, Switzerland, 2001

[ITU01b] International Telecommunications Union, ITU-R Recommendation BS. 1514, *System for Digital Sound Broadcasting in the Broadcasting Bands Below 30 MHz*, Geneva, Switzerland, 2001

[ITU02] International Telecommunications Union, ITU-R Recommendation BT. 1368-3, *Planning Criteria for Digital Terrestrial Television Services in the VHF/UHF Bands*, Geneva, Switzerland, 2002

[ITU03] International Telecommunications Union, *World Radiocommunication Conference 2003 (WRC-03)*, www.itu.int/ITU-R/go/WRC/en/, 2003

[ITU04] International Telecommunications Union, *Radio Regulations*, www.itu.int/pub/R-REG-RR/en, 2004

[ITU04a] International Telecommunications Union, *Regional Radiocommunication Conference (RRC-04)*, Geneva, Switzerland, 2004

[ITU06] International Telecommunications Union, *Final Acts of the Regional Radiocommunication Conference for Planning of the Digital Terrestrial Broadcasting Service in Parts of Regions 1 and 3, in the Frequency Bands 174–230 MHz and 470–862 MHz (RRC-06)*, Geneva, Switzerland, 2006

[ITU07] International Telecommunications Union, www.itu.int, 2007

[ITU07a] International Telecommunications Union, www.itu.int/ITU-R/terrestrial/broadcast/plans/ge06/index.html, 2007

[ITU07b] International Telecommunications Union, *World Radiocommunication Conference 2007 (WRC-07)*, www.itu.int/ITU-R/go/WRC/en/, 2007

[Jun98] Jungnickel, D., *Graphs, Networks and Algorithms*, Springer Verlag, Berlin, Germany, 1998

[Kam92] Kammeyer, K. D., *Nachrichtenübertragung*, Teubner-Verlag, Stuttgart, Germany, 1992

[Kat89] Kathrein Werke KG, *FM- und TV-Sendesysteme*, Rosenheim, Germany, 1989

[Kle96] Kleiman, S., Shah, D., Smaalders, B., *Programming with Threads*, SunSoft Press, Mountain View, California, USA, 1996

[Kor05] Kornfeld, M, Reimers, U., *DVB-H - The emerging standard for mobile data communication*, EBU Technical Review, **301** (2005), Geneva, Switzerland

[Kue98] Küchen, F., Haaß, U., Wiesbeck, W., *Vergleichende Beurteilung der Versorgungskriterien für DAB und DVB-T*, Rundfunktechnische Mitteilungen, Institut für Rundfunktechnik GmbH **42** (1998)

[Lau96] Lauterbach, T., *Digital Audio Broadcasting - Grundlagen, Anwendungen und Einführung von DAB*, Franzis-Verlag, Feldkirchen, Germany, 1996

[Leb92] Lebherz, M., Wiesbeck, W., Krank, W., *A Versatile Wave Propagation Model for the VHF/UHF Range Considering Three Dimensional Terrain*, IEEE Trans. Antennas and Propagation **40** (1992) 1121

[Lew96] Lewis, B., Berg, D. J., *Threads Primer - A Guide to Multithread Programming*, SunSoft Press, Mountain View, California, USA, 1996

[Lon86] Longley, A. G., Rice, P. L., *Prediction of Tropospheric Radio Transmission Loss over Irregular Terrain, a Computer Method*, ESSA Tech. Report ERL-79-ITS-67, Institute of Telecommunications Sciences, Boulder, USA, 1968

[LTE08] Long Term Evolution (LTE), www.3gpp.org/Highlights/ LTE/LTE.htm, 2008

[Lue84] Luebbers, R. J., *Finite Conductivity Uniform GTD versus Knife Edge Diffraction Loss in Prediction of Propagation Path Loss*, IEEE Trans. Antennas and Propagation **32** (1984) 70

[Mat83] Matula, D. W., Beck, L. L., *Smallest-Last Ordering and Clus-
         tering and Graph Coloring Algorithms*, Journal Association of
         the Computing Machinery **30** (1983) 417

[Mee83] Meeks, M. L., *VHF Propagation over Hilly Forested Terrain*,
         IEEE Trans. Antennas and Propagation **14** (1983) 480

[Mel06] Meltzer, S., Moser, G., *HE ACC v2 – Audio Coding for Today's
         Digital Media World*, EBU Technical Review, **305** (2006),
         Geneva, Switzerland

[OLe98] O'Leary, T., *Wiesbaden '95 Revisited - T-DAB Planning
         Parameters, Reference Networks and Frequency Planning
         Algorithms*, EBU Technical Review, **278** (1998), Geneva,
         Switzerland

[OLe06] O'Leary, T., Puigrefagut, E., Sami, W., *GE06 — Overview
         of the Second Session (RRC-06) and the Main Features for
         Broadcasters*, EBU Technical Review, **308** (2006), Geneva,
         Switzerland

[Oku68] Okumura, Y., Ohmori, E., Kawano, T., and Fukuda, K., *Field
         Strenght and its Variability in VHF and UHF Land-Mobile Ra-
         dio Service*, Review of the Electrical Communication Labara-
         tory **16** (1968) 825

[Phi95] Philipp, J., *Einfache und genaue Approximation der Summen-
         feldstärke log-normal verteilter Einzelfelder durch sukzessive
         Verknüpfung der Komponenten*, Kleinheubacher Berichte **38**
         (1995) 195

[Pro89] Proakis, J. G., *Digital Communications*, McGraw-Hill, New
         York, USA, 1989

[Pui04] Puigrefagut, E., O'Leary, T., *RRC-04/06 – An Overview of the
         First Session (RRC-04)*, EBU Technical Review, **300** (2004),
         Geneva, Switzerland

[Qua07] QUALCOMM, www.qualcomm.com, 2007

[Ree60] Reed, I., Solomon, I., Solomon, G., *Polynomial Codes Over
         Certain Finite Fields*, Journal of the Society for Industrial and
         Applied Mathematics. [SIAM J.], **8** (1960) 300

[Rei01]   Reimers, U., *Digital Video Broadcasting – The International Standard for Digital Television*, Springer-Verlag, Berlin, Germany, 2001

[RSD02]   Radio Spectrum Decision, *Decision No 676/2002/EC of the European Parliament and of the Council of 7 March 2002 on a Regulatory Framework for Radio Spectrum Policy in the European Community (Radio Spectrum Decision)*, Official Journal of the European Communities, (2002) L108

[RSP07]   Radio Spectrum Policy Group, *Opinion on EU Spectrum Policy Implications of the Digital Dividend*, Brussels, rspg.groups.eu.int, 2007

[She95]   Shelswell, P., *The COFDM Modulation System: The Heart of Digital Audio Broadcasting*, Electronic & Communication Engineering Journal, (1995) 127

[Sto98]   Stott, J., *The how and why of COFDM*, EBU Technical Review, **278** (1998), Geneva, Switzerland

[UPU07]   Universal Post Union, www.upu.int, 2007

[Vit67]   Viterbi, A.J., *Error Bounds for Convolutional Codes and an Asymptotically Optimum Decoding Algorithm*, IEEE Transactions on Information Theory, **13** (1967) 260

[Wei98]   Weißenfels, G., *DAB-Blockzuweisung – Lösungsansätze*, Messung und Modellierung von Funkanälen, Projekttag des Forschungsverbundes Medientechnik Südwest, Beutler, R., Prosch. Th. (Hrgs.), Stuttgart, Germany, 1998

[WiM08]   WiMAX Forum, www.wimaxforum.org, 2008

[Woo04]   Wood, D., *High Definition for Europe – An progressice approach*, EBU Technical Review, **300** (2006), Geneva, Switzerland

[Wor00]   WorldDAB, *Can We Satisfy the Future DAB Spectrum Demand*, SLC Report on DAB Spectrum Demand, London, UK, 2000

[Wor07] WorldDMB, *DAB+ – The Additional Audio Codec in DAB*, www.worlddab.org/pdf/DAB+brochure.pdf, 2007

[ZIB01] Konrad-Zuse-Zentrum für Informationstechnik, *FAP web – A website about Frequency Assignment Problems*, Berlin, fap.zib.de, 2001

[Zwi06] Zwicker, E., Fastl, H., *Psychoacoustics: Facts and Models*, Springer-Verlag, Berlin, Germany, 2006

List of Figures

List of Tables

Index